Climate Change, Aquatic Ecosystems, and Fishes in the Rocky Mountain West: Implications and Alternatives for Management

Bruce E. Rieman and Daniel J. Isaak

Rieman, Bruce E.; Isaak, Daniel J. 2010. **Climate change, aquatic ecosystems, and fishes in the Rocky Mountain West: implications and alternatives for management.** Gen. Tech. Rep. RMRS-GTR-250. Fort Collins, CO: U.S. Department of Agriculture, Forest Service, Rocky Mountain Research Station. 46 p.

ABSTRACT

Anthropogenic climate change is rapidly altering aquatic ecosystems across the Rocky Mountain West and may detrimentally impact populations of sensitive species that are often the focus of conservation efforts. The objective of this report is to synthesize a growing literature on these topics to address the following questions: (1) *What is changing* in climate and related physical/hydrological processes that may influence aquatic species and their habitats? (2) *What are the implications* for fish populations, aquatic communities, and related conservation values? (3) *What can we do about it?* In many instances, proactive efforts may help populations adapt to climate change; but elsewhere, transitions of aquatic ecosystems to alternative states may need to be facilitated. The magnitude of the challenges posed by climate change makes collaborative efforts essential among resource disciplines, agencies, and the public.

Keywords: climate change, native fishes, fisheries, stream flow, temperature, management, prioritization, resilience, vulnerability

AUTHORS

Bruce Rieman is a Research Fisheries Scientist (emeritus). He retired in 2007 from the Rocky Mountain Research Station Boise Aquatic Sciences Laboratory in Boise, Idaho, but remains active in research, management, and outreach. He holds B.S., M.S., and Ph.D. degrees from the University of Idaho. He has worked with issues in native fish conservation and fisheries management for the past 37 years.

Dan Isaak is a Research Fisheries Scientist with the Rocky Mountain Research Station Boise Aquatic Sciences Laboratory in Boise, Idaho. He holds a B.S. degree from South Dakota State University, an M.S. degree from the University of Idaho, and a Ph.D. from the University of Wyoming. He has worked in native fish conservation and fisheries management and research in the western United States for the past 18 years, and his primary research interest is in understanding how climate change, disturbance, and biophysical interactions across spatial and temporal scales affect population dynamics and habitat in headwater streams.

ACKNOWLEDGMENTS

Support for this project was provided through funding from the Rocky Mountain Research Station and the R1-R4-RMRS technology transfer program. Kerry Overton was instrumental in guiding the effort. Dona Horan provided support throughout preparation of the figures and document, and Sherry Wollrab and Laura Anglin provided editorial assistance. The final draft was improved through constructive comments provided by Kerry Overton, Kurt Fausch, James Roberts, Katherine Smith, and John Chatel.

Cover: Background: Photo courtesy of Boise National Forest. *Inset images (top to bottom):* Image courtesy of NASA Goodard Institute for Space Studies; Photo courtesy of Bart Gamett, USDA Forest Service; Image from Rieman and others (2007); Photo courtesy of Dan Isaak, USDA Forest Service.

You may order additional copies of this publication by sending your mailing information in label form through one of the following media. Please specify the publication title and series number.

Publishing Services

Telephone	(970) 498-1392
FAX	(970) 498-1122
E-mail	rschneider@fs.fed.us
Web site	http://www.fs.fed.us/rm/publications
Mailing address	Publications Distribution
	Rocky Mountain Research Station
	240 West Prospect Road
	Fort Collins, CO 80526

Executive Summary

Little question exists that Earth's climate is changing and that human causes are fundamentally important. Climate model projections indicate that the climate will continue to warm at rates equal to or faster than rates in recent decades until at least the middle of the twenty-first century. The effects of climate change could be particularly profound for native fishes and aquatic ecosystems of the Rocky Mountains because those systems often lack resilience and are strongly dependent on temperature and stream flow regimes that are already documented to be changing. The vulnerability of fish populations and communities to climate change will vary across the region based on local conditions and the amount of change that occurs. Because management budgets are limited, it will be important to prioritize limited resources based on some understanding of that vulnerability and the range of management alternatives. To consider management in the face of climate change, we have synthesized information on climate change and native fishes, stream habitats, and the observed and anticipated effects of climate change in the Rocky Mountain West. This report is organized around the following questions: *What is changing, what are the implications for native fishes*, and *what can we do about it?*

Important changes for fishes and their habitats will be driven by two factors that are the principle components of climate: air temperature and precipitation. Air temperatures across the Rocky Mountain West are warming faster than global averages and have increased by about 1 °C over the last century. Changes in precipitation are less consistent, but slightly drier summers and wetter winters are anticipated in the northern Rocky Mountains while the southern Rockies will probably experience generally drier conditions. The temporal and spatial patterns of change are not likely to be constant or linear but will vary with local trends and shorter-term climate cycles such as the Pacific Decadal Oscillation and the El Niño-Southern Oscillation. As climate change progresses, however, long-term warming trends and increasing variability will result in more frequent events (e.g., droughts, intense precipitation, and periods of unusually warm weather) that were considered extreme during the twentieth century, and the magnitude of these events may also exceed recent historical levels. Changes in stream environments will parallel trends in the climate system, with streams becoming warmer, more variable in flow timing and amount, and subject to more frequent extreme events that could be synchronized across broader areas through regional flooding, droughts, and wildfires. Climate change is also likely to influence channel structure and forest and riparian communities through altered patterns and severity or intensity of wildfire, inputs of sediment and large wood, and disturbances such as debris flows. Although stream changes have been and are anticipated to be widespread, changes will not necessarily occur at the same rates across the Rocky Mountain West. Some stream systems are changing more rapidly than others, and some even show trends opposite to general expectations. The characteristics of watersheds and streams that may either aggravate or confer some resistance to the general effects of climate change will be an important question for further research.

The implications of climate change for native fishes are challenging to anticipate because of a general lack of long-term monitoring data for Rocky Mountain populations and the complexity of interactions between biological and physical processes that are involved. In general, however, suitable habitats that are defined by temperature, flow, and other physical or biotic conditions must shift in location (generally to a higher elevation or latitude). Some species, populations, and communities may be able to track these changes and simply "relocate" but upstream limits of available habitat and barriers to dispersal and migration will limit many others. In other cases, the interaction of climate change, heterogeneous landscape responses, shifting species distributions, and new biotic and physical interactions will create novel environments, trophic cascades, and communities that have no natural precedent. These complex interactions will make prediction and management even more difficult. Natural selection and phenotypic plasticity could help many species and populations adapt to new environments, but the speed and capacity for adaptation of most species are not well known and may be outpaced by the rate of climate change.

What we do about climate change will depend on available information, management resources, and policy direction. The basic decision, however, is to either work to conserve existing species, populations, and communities through *adaptation* strategies or to *facilitate* their transitions to new conditions that are most desirable and feasible in the future. Our recommendations to guide management responses fall in five general areas. First, if the goal is **adaptation** and conservation of existing species or communities, efforts to enhance *resistance* and *resilience* of existing populations will be key. Important steps will include efforts to reduce non-climate stresses that may influence survival, growth, and habitat capacities; conserve and expand critical habitats; reconnect streams; and conserve genetic and phenotypic diversity. Second, because the potential threats of climate change and the feasibility of successful adaptation will vary widely across managed landscapes, it will be important to **prioritize limited resources**. Steps to that end include efforts to: clarify goals and values; focus on populations as fundamental units of conservation; consider the relative vulnerability and relative value among populations and habitats; and favor actions robust to uncertainty. The latter step has also been called a "no regrets" strategy, where the focus is on actions that will be useful whether the climate changes as anticipated or not. Third, if it is not likely that existing populations or communities can be maintained in the face of climate change, managers may consider efforts to **facilitate transitions to new conditions**. Facilitation might occur passively through the simple removal of existing barriers, allowing native or non-native species to move into new areas. It might be active through the intentional introduction of species to new environments. The latter option can be particularly controversial and requires clear articulation of the conservation values at stake. Fourth, the effectiveness of any action associated with the first three recommendations depends on the quality of available information to guide decisions. For that reason, it is important to **develop local information**. Important steps toward that end include efforts to: understand context based on existing climate projections and models of hydrologic, temperature, or biological changes near or encompassing the areas of interest; synthesize existing information for the area of concern such as inventory and monitoring data or local climatic trends that could be used to understand the status, distribution, and vulnerability of important species, habitats, or watersheds; model to fill gaps where useful data are limited or future projections are needed; and monitor to document trends and validate models. Finally, because the issues associated with climate change will be socially and ecologically complex and because new information and tools are developing quickly, it will be important to **coordinate efforts** across disciplines, across agencies and jurisdictions, and with the public.

The resources available to managers and biologists dealing with climate change will always be limited. The challenges associated with climate change are substantial, perhaps as important as the habitat losses already imposed on aquatic ecosystems by past human actions. The urgency is confounded by large uncertainties associated with the Earth's climate trajectory and poor understanding of how broad climate trends will translate to local effects on streams and aquatic communities. Where do we start? Some have argued that public education on aquatic issues and the trade-offs we make with other values may be the single most important thing aquatic managers and biologists can do for the long-term conservation of aquatic ecosystems, biological diversity, fishes, and fisheries. Without a fundamental change in understanding and support of aquatic conservation, not much may change. We believe that local monitoring with climate change in mind is one of the most important steps biologists and managers can take in response to climate change. Monitoring data will help test hypotheses relevant to stream system responses to climate change, will provide managers with important insight to the relative utility of management actions, and may also provide a basis for making difficult decisions. Partnerships between research and management and continued monitoring and education through both local and broader collaborative efforts could promote effective adaptation to a changing climate.

CONTENTS

Introduction

There is little question that Earth's climate is changing and that human causes are fundamentally important (figure 1). Periodic assessments of the International Panel on Climate Change (IPCC 2007) synthesize the most comprehensive evidence on the nature of global climate change and its causes, scientific uncertainty, and implications for human and natural systems. A rapidly expanding literature (e.g., Abatzoglou and Redmond 2007; Barnett and others 2008; Joyce and others 2008), instrumental record (e.g., Luce and Holden 2009; Stewart and others 2005), and other resources (e.g., http://cses.washington.edu/cig/fpt/cloutlook.shtml; http://www.fs.fed.us/ccrc/) document the nature of changes in the western United States as well as potential effects on natural resources and aquatic ecosystems (e.g., Bisson 2008; ISAB 2007; Mantua and others 2009; Nelitz and others 2009).

Effects of climate change could be particularly profound for aquatic ecosystems. Warming air temperatures and changing precipitation translate to increasing water temperatures; alteration of stream hydrology; and changes in the frequency, magnitude, and extent of extreme climate events such as floods, droughts, and wildfires (Hamlet and Lettenmaier 2007; Jentsch and others 2007; McKenzie and others 2004). The biology of aquatic organisms is largely dependent on temperature and flow and most have evolved in a dynamic environment defined by hydrologic and geomorphic processes (Bisson and others 2009; Brannon and others 2004; Waples and others 2009). However, fundamental changes in climate could lead to fundamental changes in physiology, behavior, and growth of individuals (Jager and others 1999; Neuheimer and Taggart 2007; Pörtner and Farrell 2008); phenology, growth, dynamics, and distribution of populations (Hari and others 2006; Rieman and others 2007; Robards and Quinn 2002); persistence of species and structure of communities (Finney and others 2002; Hilborn and others 2003); and functioning of whole ecosystems (Moore and others 2009).

In relation to environmental variation over evolutionary and ecological time scales that shaped existing species of fishes and phenotypic diversity within those species (approximately 10^3 to 10^6 years, respectively; Lichatowich 1999; Waples and others 2009; Wood and others 2008), contemporary changes in climate may be relatively minor. Indeed, many species such as the Pacific salmon, trouts, and chars that are now a focus of intensive conservation management have recovered, diversified, and expanded in distributions following continental glaciation, volcanism, and the cataclysmic processes that shape mountain streams and rivers (Chatters and others 1995; Montgomery 2000; Wood and others 2008). The effects of contemporary climate change, in contrast to the degradation and loss of aquatic habitats associated with human development over the

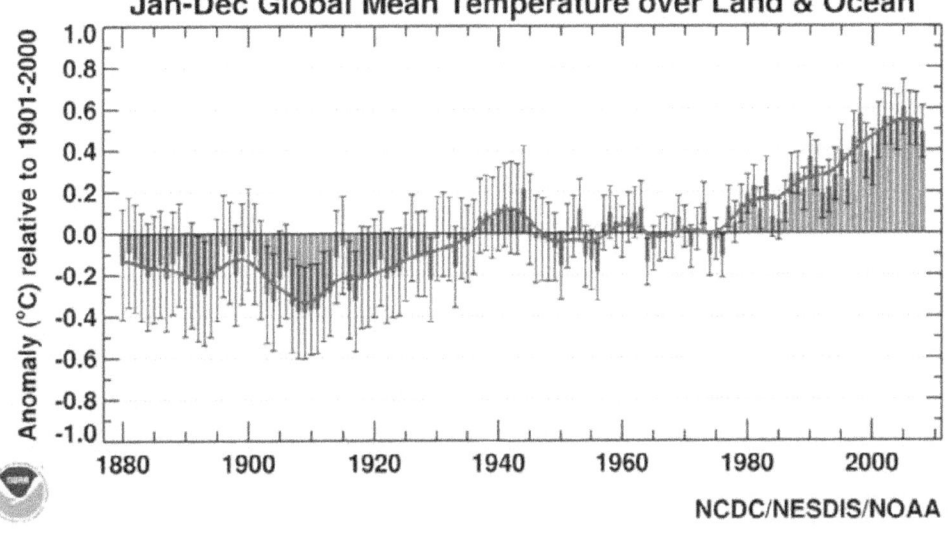

Figure 1. Instrumental record of average global air temperatures from 1880 to 2008 (from NASA Goddard Institute for Space Studies).

last two centuries, might represent a relatively minor loss of historical habitat capacity or productivity. The problem is that modern climate change is occurring especially quickly at the end of an already warm period (IPCC 2007; Lüthi and others 2008), indicating a potential for new conditions with no natural precedent (Williams and others 2007), and in the wake of already extensive habitat and aquatic community disruption (Wood and others 2008). Many species and populations no longer have the networks and diversity of habitats and refugia, genetic diversity, or evolutionary potential that allowed them to resist or rebound in the face of past environmental disturbance and change. In some cases, anticipated changes may outpace the remaining capacity for adaptation and dispersal (Crozier and others 2008).

Vulnerability of aquatic species, populations, and communities to climate change will depend on a context defined by the characteristics of those species and local environments, past habitat disruption, fragmentation and loss, and the nature of the change that occurs. As a result, that vulnerability may vary dramatically across populations, species, and landscapes that are a focus of land and natural resource management (e.g., Rieman and others 2007). In most cases, capacity for conservation management is constrained by limited budgets, time, and other resources. It is also constrained by current understanding of the implications of climate change and management actions or alternatives that might effectively influence results. It will be important to prioritize limited resources and to guide management based on some understanding of the vulnerability of species, populations, and ecosystems of interest.

The objectives of this report are to synthesize information on native fishes, stream habitats, and the anticipated effects of climate change in the Rocky Mountain West. This region may be particularly sensitive to climate change given the rate at which it is warming and because many of its ecosystems are constrained by water availability (Brown and others 2008; Saunders and others 2008). It is not our intent to provide an exhaustive review of the climate-aquatic-fisheries literature; several good ones already exist (Bisson 2008; Ficke and others 2007; Francis and Sibley 1991; Furniss and others 2010; Hauer and others 1997; Heino and others 2009; ISAB 2007; Poff and others 2002; Rahel and others 2008; Schindler and others 2008). Our intent is to provide an overview of important information as context for management that might begin to address the implications of climate change. The report is organized around the following questions:

- *What is changing* in climate and related physical/hydrologic processes that may influence aquatic species and their habitats?

- *What are the implications* for fish populations, aquatic communities, and related conservation values?

- *What can we do about it?*

What Is Changing?

Climate

Air temperature and precipitation are the principle components that constitute "climate." Short-term variability in these factors is often considered to be the "weather" and seasonal cycling, whereas long-term variation (i.e., decadal to millennial timescales) constitutes climatic regimes (figures 1 and 2). Interactions among climate, geology, and topography have formed the distinctive physiographic regions into which species of the Rocky Mountains have dispersed and evolved over millennial timescales (Hessburg and others 2005; McPhail and Lindsey 1986; Schumm and Lichty 1965; Wolock and others 2004). The expansive and varied regions across the Rocky Mountain West are characterized by a diversity of climatic conditions, ranging from greater than 2000 mm of annual precipitation and mean air temperatures of approximately 0 °C at the highest elevations in the north to less than 100 mm of annual precipitation and annual temperatures exceeding 15 °C in parts of the southern Rockies (Hijmans and others 2005). Annual temperature cycles create strong seasonality in most northern areas, and precipitation during winter often occurs as snow and accumulates at higher elevations until warming spring temperatures drive snowmelt and broadly synchronized patterns of stream runoff (Barnett and others 2008; Stewart and others 2005). This general pattern holds only at high elevations in the southern interior due to latitudinal clines toward warmer temperatures and decreases in snow accumulation. At lower elevations, seasonal cycles are less pronounced and stream flows are more variable, with many smaller streams exhibiting periods of drying interspersed with periodic flooding from summer monsoons and convective thunderstorms (Minckley and Deacon 1991; Propst and others 2008).

Climate has been changing across the Rocky Mountains in association with global patterns. However, instrumental records suggest that mean annual air temperatures during the twentieth century increased by approximately 1 °C (Saunders and others 2008), which is considerably more than the 0.6 °C increase in global temperatures for the same period (IPCC 2007). The larger increase is due to warming rates that are faster over land masses than over the oceans (the global average includes both), but the Rocky Mountain West also has been warming more rapidly than other areas of the coterminous United States (Saunders and others 2008). This spatial heterogeneity is also apparent at finer scales across the interior, where a four-fold variation in warming rate has been observed and a few locations even exhibit cooling trends (figure 3a; Mote and others 2005). Adding further complexity is seasonal variation in warming where trends have generally been most pronounced in the spring, early summer, and winter and have been subdued in the late summer and fall (Abatzoglou and Redmond 2007; Knowles and others 2006; Pederson and others 2009).

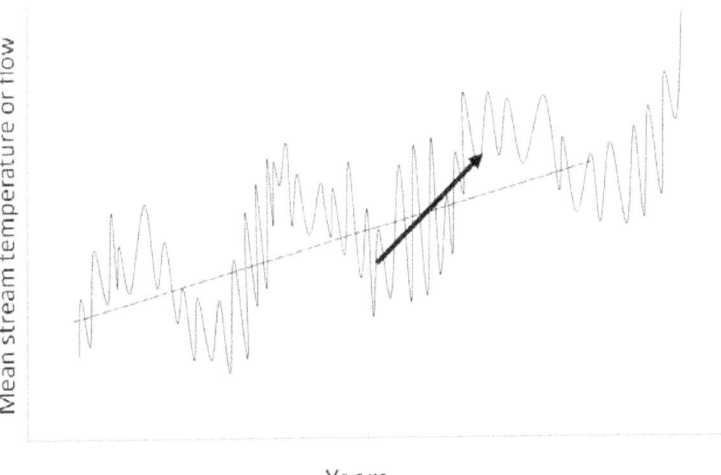

Figure 2. Stream temperature, flow, and other conditions influenced by climate vary from year to year and over longer periods associated with periodic phenomena like the Pacific Decadal Oscillation and the El Niño-Southern Oscillation. The overall trends associated with climate change are evident in long-term monitoring data, but trends over shorter periods (e.g., the black arrow in the figure) may be more strongly positive, neutral, or even negative and obscure the longer-term patterns for periods of time.

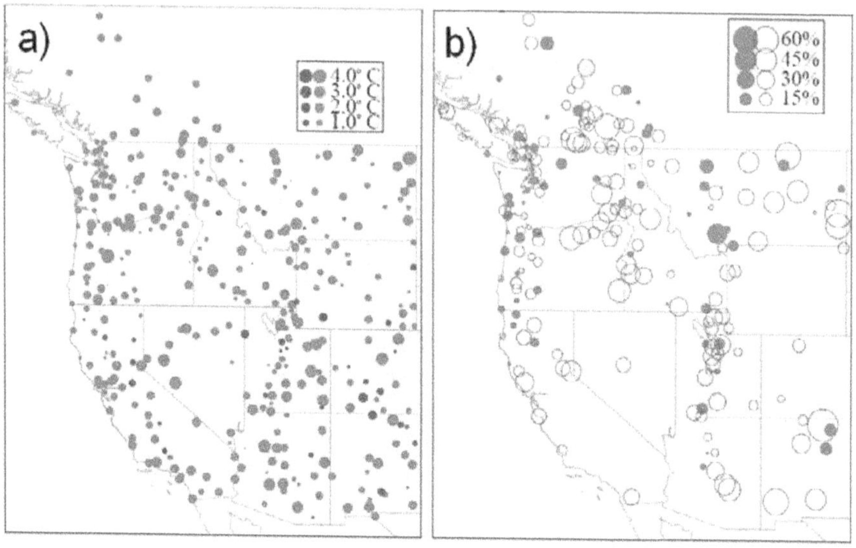

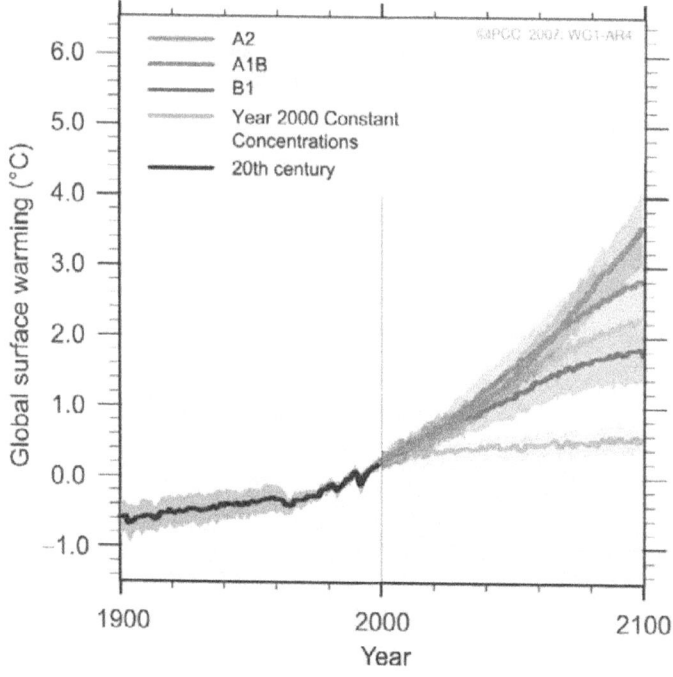

Figure 3. Observed spatial variability in (a) air temperature and (b) precipitation trends across the western United States from 1930 to 1997 (from Mote and others 2005). Red (blue) circles indicate warming (cooling) air temperatures or decreasing (increasing) precipitation.

Trends in total annual precipitation during the twentieth century are less consistent than those of air temperatures, although small increases in precipitation have been observed in many areas of the Rocky Mountains (figure 3b; Hamlet and others 2007; Mote and others 2005), and changes in warm season/cool season proportions are apparent in some regions (Hamlet and others 2007). More importantly, perhaps, is that a larger proportion of precipitation now falls in extreme events (Wehner 2005) and as rain rather than snow due to the interaction with warmer air temperatures (Knowles and others 2006).

As substantial as the trends in twentieth century climate have already been, Global Climate Models (GCM) project continuation and even acceleration of warming through at least the middle of the twenty-first century (figure 4; IPCC 2007). Current projections for the western United States suggest mean annual air temperatures will increase by another 1 °C to 3 °C by mid-century (Mote and others 2008; GCRP 2009), and early indications put the Earth's climate trajectory toward the higher end of this range (Pittock 2006; Raupach and others 2007). Even the most conservative estimates suggest a warming rate in future decades that is twice that

Figure 4. Global climate change scenarios based on different assumptions regarding greenhouse gas emissions and trends in human societies (from IPCC 2007). Note that most scenarios predict similar amounts of warming in the middle of the twenty-first century, but large differences emerge later in the century due to unknowns about greenhouse gas emissions in the next few decades.

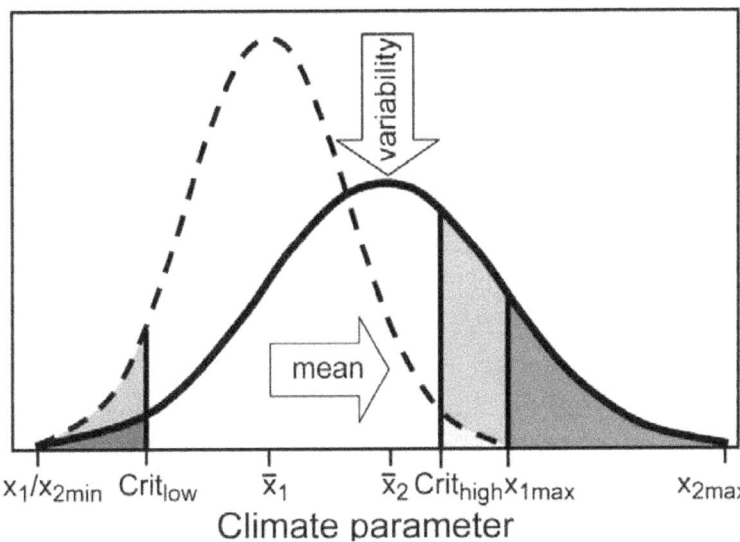

Climate parameter

x₁/x₂min Crit_low x̄₁ x̄₂ Crit_high x₁max x₂max

Figure 5. Changes in the probability of extreme weather events (red and blue distribution tails) associated with a shift in the mean climate and increased variability (from Jentsch and others 2007).

observed during the twentieth century. Moreover, these increases in temperature are expected to be accompanied by greater overall variability. Thus, it may still be possible to continue exceeding historical records for cold temperatures even as record highs become increasingly frequent (Meehl and others 2009). More importantly, the combination of a shift toward higher mean temperatures and enhanced variability implies that the frequency of weather conditions now considered "extreme" could increase rapidly (figure 5; IPCC 2007; Jentsch and others 2007).

Projections regarding changes in precipitation are generally less certain than those for air temperatures. Most GCMs forecast small precipitation increases for the northern interior, accompanied by a seasonal shift toward drier summers and wetter winters (Mote and others 2008; GCRP 2009). In the south, however, significant annual precipitation decreases on the order of 15 to 40 percent are projected, and this area is one of the few for which GCM projections have a high level of agreement (figure 6; Hoerling and Eischeid 2007; GCRP 2009).

Superimposed on climate warming trends are cycles associated with the Pacific Decadal Oscillation (PDO), Atlantic Multidecadal Oscillation (AMO), and the El Niño-Southern Oscillation (ENSO) that periodically enhance and dampen long-term trends (figures 2 and 7). All

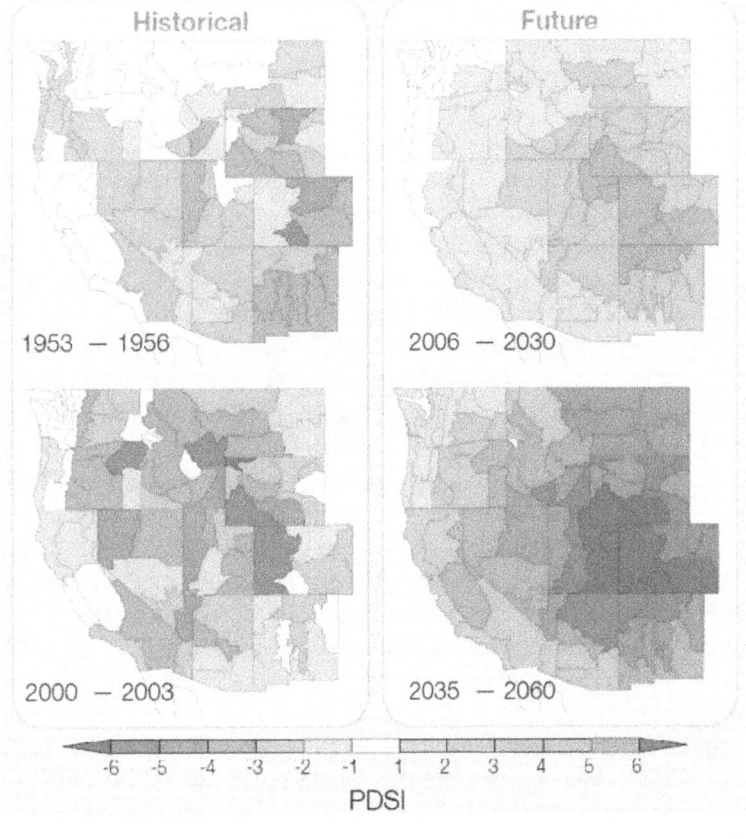

Figure 6. Distribution of Palmer Drought Severity Index scores based on historical conditions and future projections (from Hoerling and Eischeid 2007).

such climate cycles are derivatives of atmospheric-oceanic heat exchanges and manifest as multi-year periods with above (or below) average temperatures and precipitation across the western United States (McCabe and others 2004; McPhaden and others 1998;

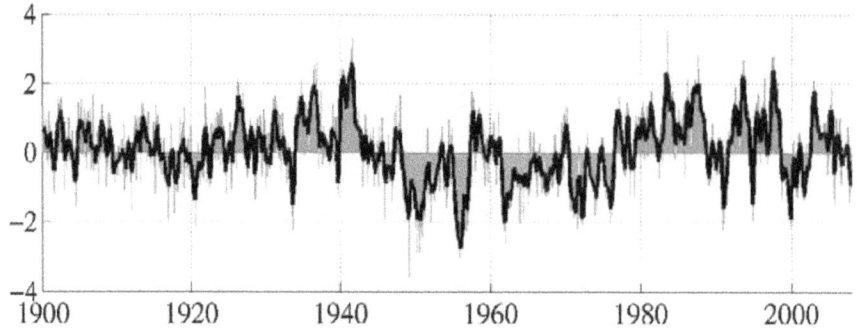

Figure 7. Monthly values of the Pacific Decadal Oscillation from 1900 to 2008 (N. Mantua, unpublished data; http://jisao.washington.edu/pdo/).

Mote and others 2003). The length, consistency, and intensity of cycling depend on how the phases (cool and wet or warm and dry) are aligned and vary with periodicities of the oscillations (ENSO is approximately 2 to 7 years; PDO and AMO are approximately 20 to 30 years; Mantua and others 1997; McPhaden and others 1998). There is some evidence for more frequent and intense cycling with especially warm climatic periods in the Earth's distant past (Wara and others 2005), and similar changes could occur in the future (Yeh and others 2009).

Although considerable uncertainty exists regarding future climate, it is safe to assume that managers in the Rocky Mountains will have to adjust to a warmer and sometimes drier world. Continued warming is a virtual certainty given the greenhouse gases already emitted, inertia in the global climate system, and human energy demand. But just how different the future will be is difficult to predict. A best case scenario is temperature gains comparable to those already experienced during the twentieth century but compressed into approximately half the time with a transition to a new climate equilibrium sometime in the mid to late twenty-first century. Other (potentially more likely) scenarios include greater warming over a longer period, with higher temperatures exacerbated in some areas by decreases in precipitation and growing human populations with water supply needs. Moreover, the long life of greenhouse gases in the atmosphere means that even after emissions are curbed, it may take several millennia for reversion to a cooler climate that would approximate recent historic conditions (Solomon and others 2009). Future environmental changes, therefore, might be considered more or less permanent with regard to any practical management horizons.

Stream Environments

Channel and riparian conditions. Climate change during the twentieth century resulted in a series of related environmental trends across western North America that could translate to wide-ranging effects on channel structure and riparian conditions (van Mantgem and others 2009; Yarnell and others 2010). Warmer air temperatures, for example, have decreased winter snow accumulations by increasing the rainfall fraction, advancing spring snowmelt, and altering the pattern of stream flow recession (Hamlet and others 2005; Mote and others 2005; Yarnell and others 2010). Smaller snowpacks that melt sooner have translated to increasing drought frequency and severity (Hoerling and Eischeid 2007) and more extensive wildfire activity (Littell and others 2009; Morgan and others 2008; Westerling and others 2006). These stresses are in turn effecting changes in the composition of the region's forest, rangeland, and riparian plant communities, with more xeric, drought-tolerant species becoming more common (Breshears and others 2009; van Mantgem and others 2009). Changes are occurring gradually through direct climate effects on demographic processes (van Mantgem and others 2009), during dramatic regional die-offs associated with extreme droughts and wildfires (Breshears and others 2005; van Mantgem and Stephenson 2007), and by making plant communities more susceptible to invasions by non-native plants and outbreaks of pests like mountain pine beetles (*Dendroctonus ponderosae*; Logan and Powell 2001; Pettit and Naiman 2007).

Altered forest and riparian communities, combined with wildfire activity, will change inputs of sediment and large wood, and these basic channel constituents will be routed differently by hydrologic regimes that are also evolving (Barnett and others 2008; Miller and others 2003). In steep topographies, post-fire debris flows could become more common in small channels and sometimes cause extirpations of local fish populations (e.g., Bozek and Young 1994; Brown and others 2001). Debris flows may also simplify habitats in small, steep channels through scour and removal of local alluvium, bank soils, and woody debris, while paradoxically creating more diverse habitats in downstream channels where sediment and wood deposition occur (Miller and others 2003; Reeves and others 1995). At present, few studies have linked channel evolution to rapid, anthropogenic climate change in mountainous areas, but see a review by Blum and Tornqvist (2000) for information on longer-term (i.e., 10^2 to 10^4 years) channel responses that could prove relevant. With the exception of debris flows, however, time lags between alteration of hillslope processes and channel response mean that aquatic ecosystems and management options are more likely to be constrained by rapidly changing hydrologic and temperature regimes.

Hydrologic regimes. The interaction between warming temperatures and precipitation has been associated with a variety of hydrologic alterations during the twentieth century. These changes have been well documented by a monitoring network that consists of several hundred flow gages with long-term records across the western United States (e.g., Regonda and others 2005; Stewart and others 2005). Data from these sites suggest that the timing of stream runoff steadily advanced during the latter half of the twentieth century and now occurs 1 to 3 weeks earlier (figure 8; Regonda and others 2005; Stewart and others 2005), due largely to concurrent decreases in snowpack and earlier spring

melt (Mote and others 2005; Stewart and others 2005). These changes also diminish recharge of subsurface aquifers that support summer baseflows (Hamlet and others 2005), and flow declines during this period are also apparent across many Rocky Mountain streams (Rood and others 2008; Stewart and others 2005). In watersheds with densely forested vegetation, these declines may be exacerbated as a warmer climate increases water loss through evapotranspiration (Hamlet and Lettenmaier 1999; Hamlet and others 2007). In one striking example from the Pacific Northwest, Luce and Holden (2009) found that three-fourths of the 43 gage records they examined exhibited statistically significant declines in summer low flows (25th percentile or one-in-four year low flow) of 29 to 47 percent during the latter half of the twentieth century (figure 9). That decline has been accompanied by longer inter-annual correlation, indicating that when extreme droughts occur, they are more likely to persist across multiple years (McCabe and others 2004; Pagano and Garen 2005).

Mid-winter flood frequency and extent also appear to be affected by warming air temperatures, and streams in watersheds with temperatures that are near freezing are especially sensitive. Streams in these watersheds often exhibit "transitional" hydrologies that are a mix of rain

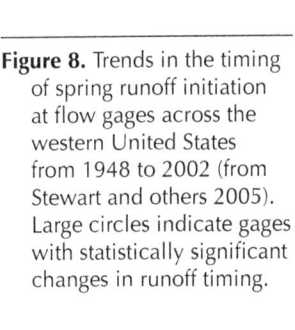

Figure 8. Trends in the timing of spring runoff initiation at flow gages across the western United States from 1948 to 2002 (from Stewart and others 2005). Large circles indicate gages with statistically significant changes in runoff timing.

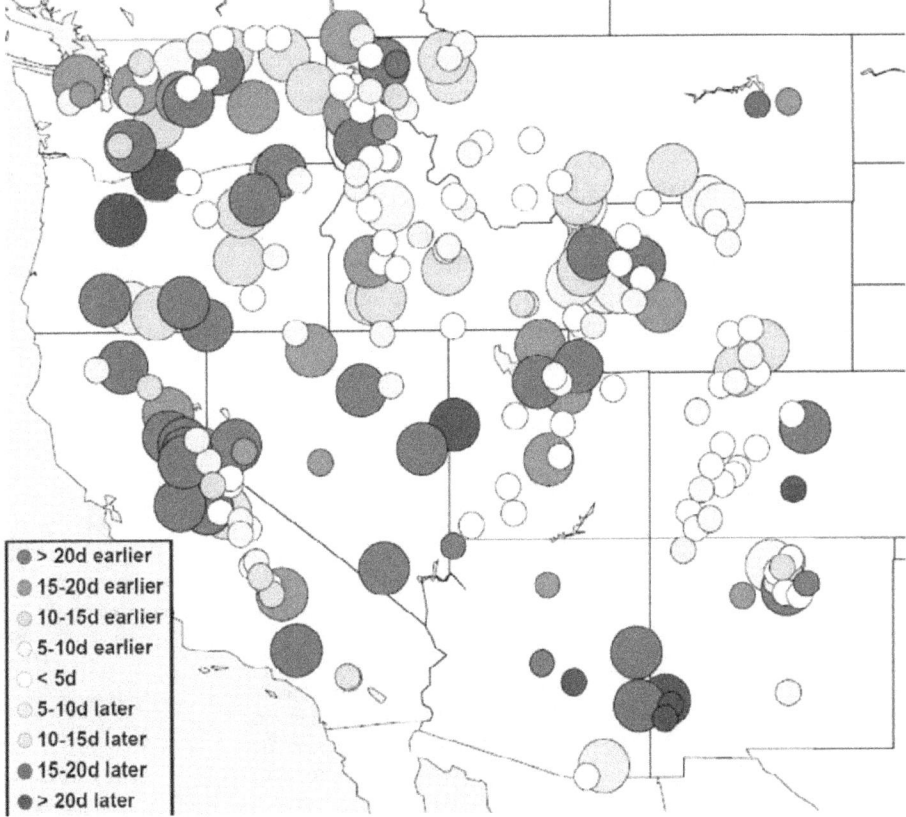

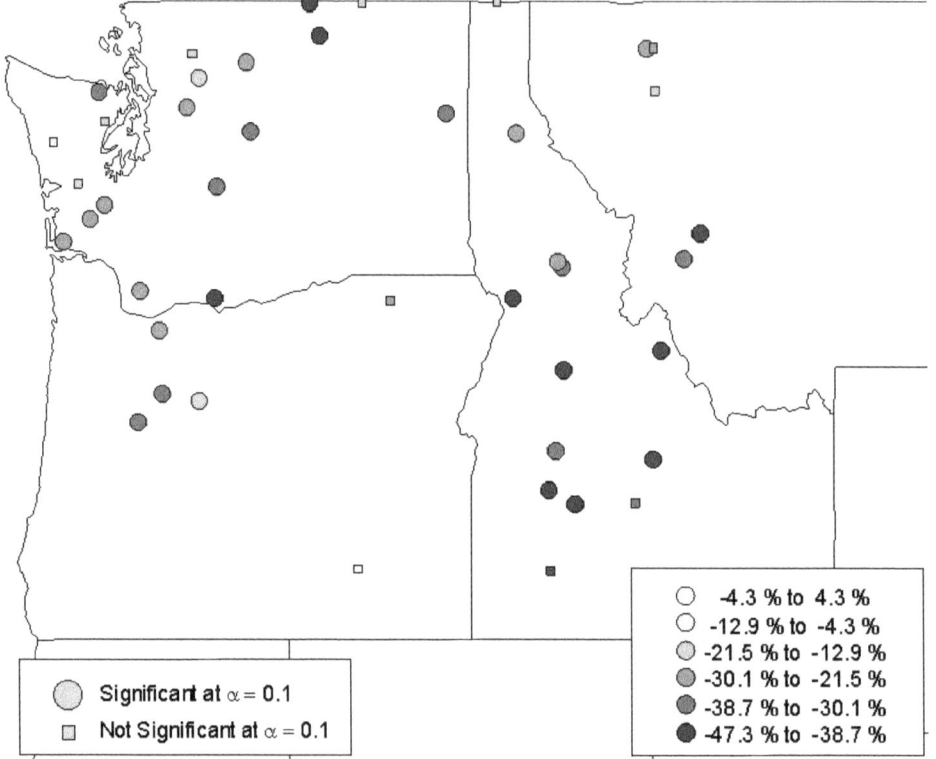

Figure 9. Trends in the 25th percentile low flows at gaging stations across the Pacific Northwest from 1948 to 2006 (from Luce and Holden 2009).

Significant at $\alpha = 0.1$

Not Significant at $\alpha = 0.1$

-4.3 % to 4.3 %
-12.9 % to -4.3 %
-21.5 % to -12.9 %
-30.1 % to -21.5 %
-38.7 % to -30.1 %
-47.3 % to -38.7 %

and snowmelt runoff. Any additional warming in these watersheds translates directly to an increase in the proportion of precipitation falling as rain and increases the potential for rain-on-snow linked to mid-winter floods (Hamlet and Lettenmaier 2007; Marks and others 1998). Streams at lower, warmer elevations where hydrology is already dominated by rainfall, and streams at higher elevations, where mid-winter temperatures will remain well below freezing for the foreseeable future, will be less sensitive (figure 10; Hamlet and Lettenmaier 2007).

Although climate change is driving systematic changes in stream hydrology, important local variability cannot be discounted. Trends toward decreasing summer flows and advances in runoff, for example, may be weaker in watersheds where local precipitation rates are trending higher, where snow is a minor source of stream flow, or where warming rates are slower (figure 3). Underlying geologies may also buffer some streams better than others against changes by providing greater subsurface aquifer storage and contributions to summer

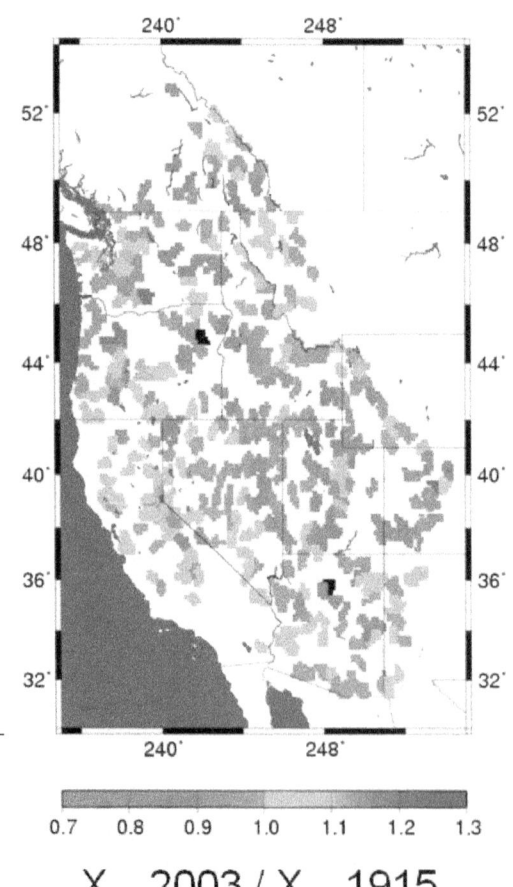

X_{20} 2003 / X_{20} 1915

Figure 10. Relative changes in 20-year flood probability from 1915 to 2003 west of the Continental Divide modeled using the Variable Infiltration Capacity hydrologic model. Watersheds in blue had increased probabilities while those in brown or red had reduced probabilities (from Hamlet and Lettenmaier 2007).

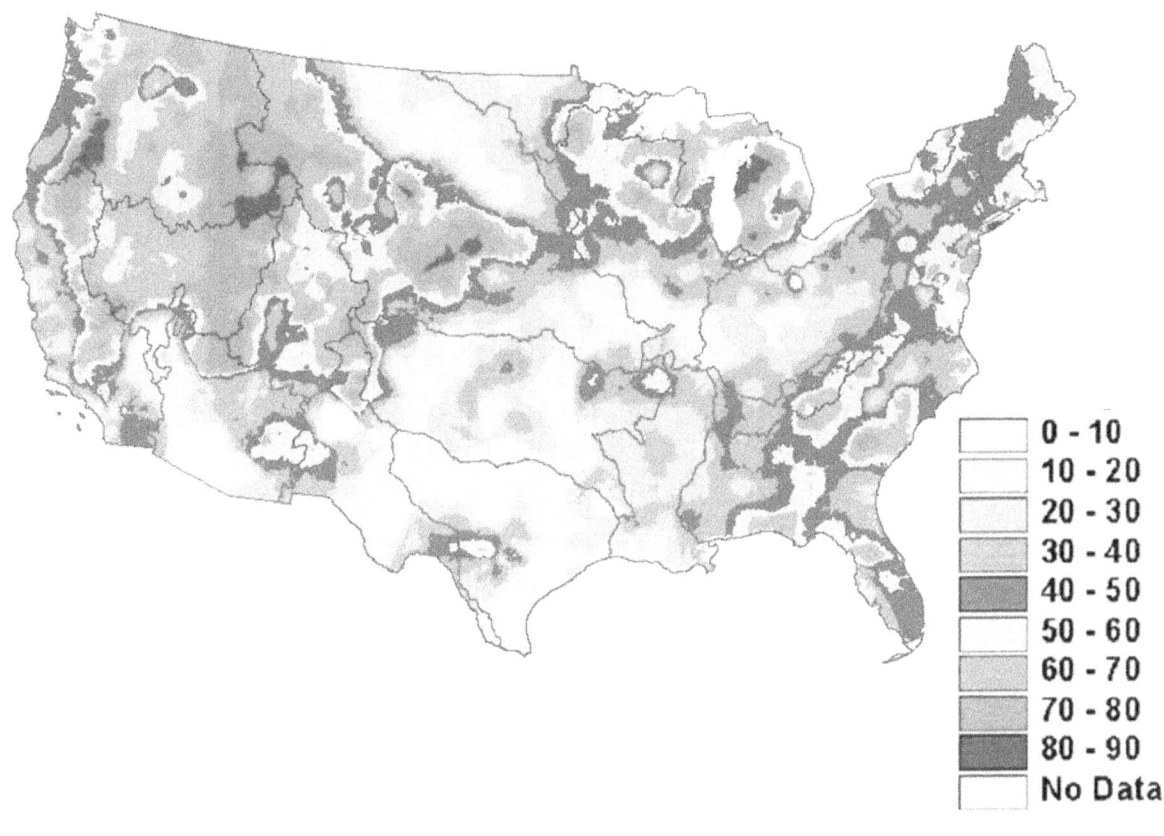

	0 - 10
	10 - 20
	20 - 30
	30 - 40
	40 - 50
	50 - 60
	60 - 70
	70 - 80
	80 - 90
	No Data

Figure 11. Baseflow index calculated as the ratio of baseflow to peakflow across the U.S. (from Santhi and others 2008). Higher values of the index are an indication of flow stability and potentially greater groundwater inputs.

baseflows. Work by Tague and others (2008) in central Oregon streams suggests this buffering effect may be substantial, with summer flows in some streams projected to decline one-third as much as others (-15 percent versus -45 percent) under different climate change scenarios. Note, however, that this study occurred in an area with pronounced differences in geology that may not be indicative of other areas that are comparable in size. Regardless, it is clear that significant variation exists across the region (figure 11; Santhi and others 2008).

Temperature regimes. Long-term temperature records from lakes and streams are relatively rare in contrast to flow records. Where long-term data are available, trends generally tracked air temperature increases in the latter half of the twentieth century (figure 12; e.g., Bartholow 2005; Morrison and others 2002; Peterson and Kitchell 2001; Robards and Quinn 2002). In one recent assessment, Kaushal and others (2010) examined river temperature trends across the United States and found that 14 of 16 western rivers had warmed during the last 30 years (10 of the warming

trends were statistically significant). The average rate of increase was 0.17 °C/decade, with the warming trend exacerbated in some places by urbanization and water development.

To compensate for a lack of long-term temperature data from lakes, Schneider and others (2009) used time-series of satellite imagery calibrated to a few years of lake surface temperature data to reconstruct previous thermal regimes. This study suggests that from 1992 to 2008, all of the lakes examined in California and Nevada showed statistically significant warming trends. Interestingly, the surface temperatures of the lakes warmed at approximately twice the rate of local air temperature increases, although it was unclear whether lake heat gains were integrated throughout the full depth profile.

Temperature data for small headwater streams are abundant, but like lakes, long-term time-series are relatively rare because inexpensive digital temperature sensors became widely available only in the 1990s (figure 13; Dunham and others 2005). To compensate for this lack of data, Isaak and others (2010) compiled a spatially extensive temperature database from multiple

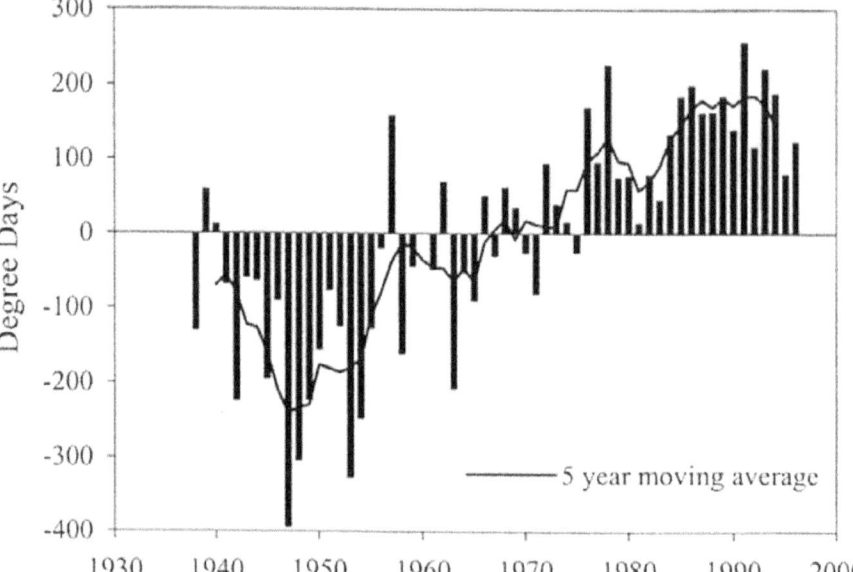

Figure 12. Trend in total annual degree days for the Columbia River at Bonneville Dam from 1938 to 1998 (Robards and Quinn 2002).

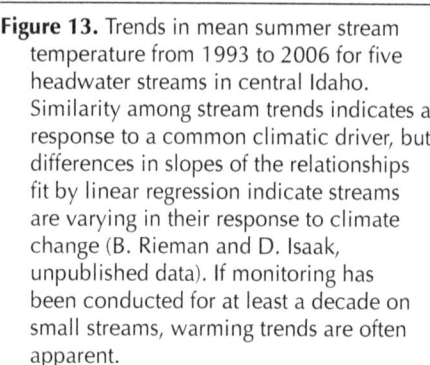

Figure 13. Trends in mean summer stream temperature from 1993 to 2006 for five headwater streams in central Idaho. Similarity among stream trends indicates a response to a common climatic driver, but differences in slopes of the relationships fit by linear regression indicate streams are varying in their response to climate change (B. Rieman and D. Isaak, unpublished data). If monitoring has been conducted for at least a decade on small streams, warming trends are often apparent.

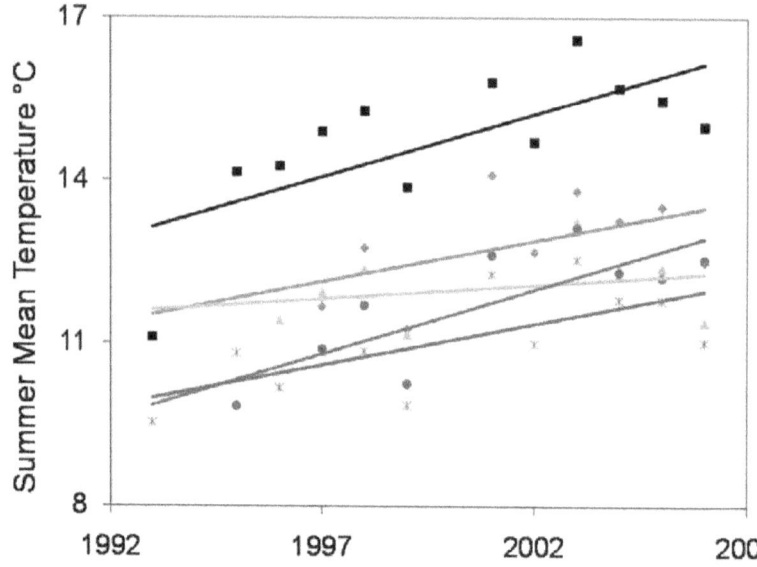

resource agencies across a large, central Idaho river network and used the data to build statistical models with climate predictors so that long-term trends could be reconstructed and examined. Results suggest that climate-related trends in air temperatures and stream flow had increased mean summer stream temperature by +0.38°C across the network from 1993 to 2006, which translated to a warming rate of 0.27 °C/decade (figure 14; Isaak and others 2010).

Another limitation of temperature data from headwater streams is that sampling has often occurred only during summer months because of difficulties with site access or maintenance of sites with large annual floods. Unfortunately, therefore, few details are known regarding how full annual thermal regimes may be changing

(e.g., timing of spring onset, changes in growing season length, and differences in seasonal trends) or the potential consequences for aquatic biotas (Olden and Naimen 2009; Danehy and others 2010). Isaak and others (2010) recently developed a simple protocol that involves gluing temperature sensors directly to large boulders that facilitates the collection of more annual stream temperature data, but it may take several years to accumulate the data necessary for assessing annual patterns.

Aside from the systematic changes driven by climate change, there are a variety of factors that will impart local variability in stream warming rates. This local variability is apparent among streams, indicated by differences in regression slopes with warming trends in figure 13. It is also illustrated by inter-annual differences at sites.

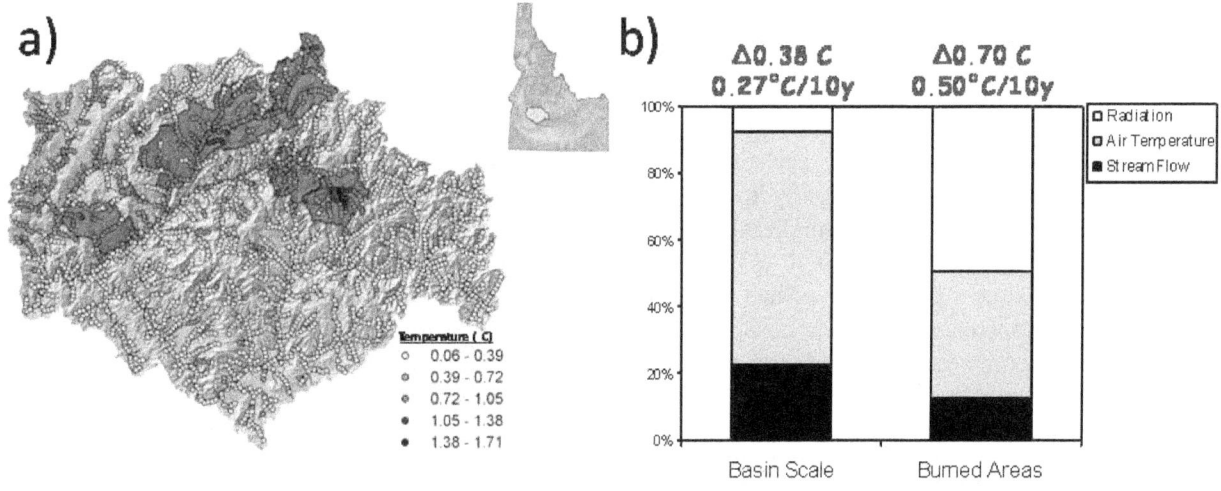

Figure 14. (a) Increases in mean summer stream temperature across the Boise River stream network from 1993 to 2006 due to wildfires and long-term trends in air temperatures and decreasing summer flows. (b) Relative influence of factors on stream temperature increases. Total stream temperature increase and decadal warming rates for the entire network and for areas within wildfire perimeters are shown above the bars (from Isaak and others 2010).

In the Boise River basin, for example, where the same 33 sites were sampled in 2007 and 2008 (figure 15), the average change in mean summer stream temperature was -1.59 °C, but a range of temperature responses was observed from -0.49 °C at the least sensitive site to -2.41 °C at the most sensitive site (the average site-level deviation from the systematic shift between years was -0.38 °C). This variable response could have been caused by variation in local microclimatic forcing (Daly and others 2009) or the sensitivity of streams to a given level of forcing (Hari and others 2006). Over longer time or broader spatial scales, differences in stream temperature response could also result from spatial heterogeneity in climate trends, changes in riparian vegetation conditions related to wildfire or drought, and evolving hydrologic regimes. The potential effects of wildfires on stream temperatures are well documented (Dunham and others 2007; Hitt 2003; Isaak and others 2010), but broad differences in hydrology may also be important (Mohseni and others 1999; Morrill and others 2005). In a national-scale assessment, Mohseni and others (1999) found that streams dominated by snowmelt runoff warmed only 0.44 °C for every 1 °C air temperature increase, in contrast to the 0.67 °C average across the United States. This result was attributed to buffering by snowmelt groundwater and implies that many streams across the Rocky Mountains may be currently less sensitive to warming. If this is true, it also implies that future sensitivity may increase as snowpacks decline.

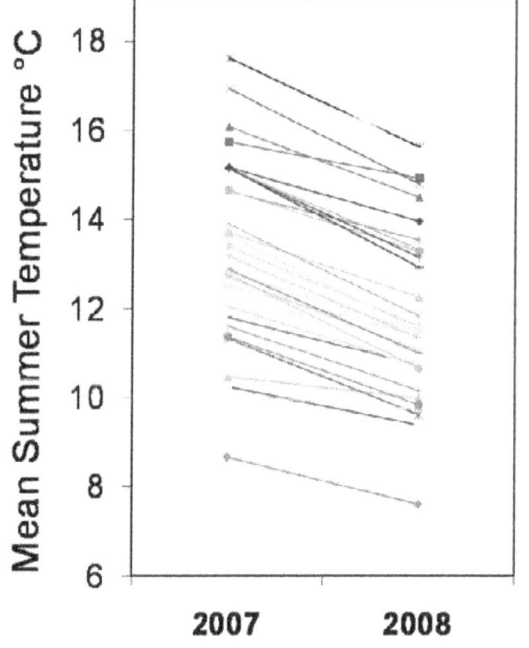

Figure 15. Variation in mean summer stream temperature changes between 2007 and 2008 at 33 sites in the Boise River network. The mean change between years was -1.59 °C (represented by the systematic change across all sites), but the average site-level deviation from the mean change was 0.38 °C (represented by the differences in slopes among all sites). These patterns indicate that although streams across broad areas may respond similarly to climate change, some streams will be more (or less) sensitive than others.

Implications for Native Fishes

The influence of climate change on stream environments, patterns of disturbance, hydrologic and geomorphic process holds important implications for native fishes and aquatic communities in Rocky Mountain streams. There is growing evidence that terrestrial ecosystems are responding to climate change through shifting distributions, invasions, local extinctions, and reorganization of whole communities and food webs (CCSP 2009; Parmesan and Yohe 2003; Root and others 2003). There is less documentation of effects in aquatic ecosystems (Heino and others 2009), but important changes have been observed in species distributions (Hari and others 2006), phenology and expression of life histories (Hilborn and others 2003; Quinn and Adams 1996; Quinn and others 1997), and basic hydrologic and limnological processes that ultimately structure populations and communities (Isaak and others 2010; Mantua and others 2009; Poff and others 2001; Schindler and others 2008; Wrona and others 2006). A slower pace of documentation in aquatic systems is not necessarily cause for optimism but perhaps a reflection of the relative difficulties of observation; complex linkages among climate, terrestrial, and aquatic systems; and more limited long-term monitoring in aquatic environments (Heino and others 2009; Isaak and others 2010). Because stream temperature and flow represent fundamental constraints on physical and biological processes in aquatic systems, climate driven changes seem almost inevitable (Crozier and others 2008; Jonsson and Jonsson 2009; Pörtner and Farrell 2008). The direct constraints imposed by thermal tolerances and stream flows are relatively easy to anticipate, and most efforts to consider the implications of climate change have focused on these effects. Interactions of physical, biological, and ecological processes will add complexity (table 1) and will lead to indirect effects that are also likely to be important but more difficult to anticipate (Jager and others 1999; Yarnell and others 2010).

Temperature and Hydrology as Primary Controls

Temperature and flow are not only very sensitive to climate forcing, but they also directly influence many of the physical and biological processes that are basic to aquatic organisms and communities (Olden and Naiman 2009; Poff and others 2010). Temperature controls solute chemistry for oxygen, CO_2, primary nutrients, trace elements, and toxic substances. Temperature and flow interact to influence mixing of water masses and create important thermal and solute gradients (e.g., Tiffan and others 2009). Temperature directly controls the rates of biochemical and physiological processes such as protein synthesis and disruption, photosynthesis, energy conversion, and respiration (Coutant 1999; Jonsson and Jonsson 2009; McCullough and others 2009). For cold blooded organisms, including most aquatic species and especially fish, temperature controls capacity for, and efficiencies of, activity, food consumption, metabolism, and growth (Coutant 1999; Pörtner and Farrell 2008). Ultimately, temperature defines the gradients of performance and absolute bounds for life for most aquatic organisms as well as rates of growth and timing of key life history events or transitions.

The influences of stream hydrology may be less constraining than temperature (many aquatic organisms, including fish, can survive in standing water), but they are still fundamentally important. Flow influences temperature (Isaak and others 2010; Meier and others 2003) and acts as a primary control on the supply and exchange of materials such as oxygen, nutrients, organic matter and food, and metabolic wastes (Harvey and others 2006; Wrona and others 2006). Flow controls the volume of habitat in streams and, through interactions with temperature and channel characteristics, the structure, distribution, and spatiotemporal availability of distinct resources such as spawning, foraging, migration, and refuge habitats that are critical to completion of life cycles (Geist and others 2008; Northcote 1997; Yarnell and others 2010). Multiple authors argue, for example, that the distribution and success of stream salmonids is fundamentally shaped by the interaction of hydrologic and temperature regimes across the range of habitats required to complete life histories and the capacity for adaptation in life history necessary to persist in that context (Brannon and others 2004; Crozier and others 2008; McClure and others 2008). Ultimately, temperature and flow interact with geology, geomorphology, and terrestrial and riparian communities to create the mosaic of habitats and environmental conditions that represent the template for species occurrences and dynamics and the expression and evolution of aquatic biological diversity (Bisson and others 2009; McClure and others 2008; Poff and others 2001).

Table 1. Examples of aquatic biological and ecological conditions and processes important in potential responses to climate change at different levels of biological organization. Although important effects are likely at all levels of biological organization, higher levels of organization are increasingly complex, leading to limited capacity to predict the direction or magnitude of changes with any certainty.

Level of organization

Conditions or processes likely to change

Molecular

Solute chemistry, rates of biochemical processes, structural constraints on proteins

Physiological

Metabolism (respiration, consumption, excretion, energy conversion)
Scope for growth and activity
Capacity for osmotic regulation

Organism

Changing rates of embryo development and timing of emergence relative to food and disturbance
Length, seasonal timing, and suitability of periods of growth or stress
Realized growth and growth-dependent life history, age of maturation, fecundity, and migration. Potential changes in size and timing for migration, growth for overwinter survival, development of novel combinations of conditions with unpredictable results
Migration success linked to physiological exhaustion with high temperature or low water migration barriers
Increasing probability of acute stresses and lethal temperature or flow conditions for some coldwater species, increasing opportunity for others

Population

Changing rates of mortality, reproduction, and recruitment and inherent population growth rates
Changing frequency of catastrophic mortalities
Shifting distribution, creation, and loss of suitable habitats; changing access to new habitats
Changing size, geometry, and connection of existing habitats

Community

Altered rates of predation, competition, parasitism, and disease
Changing forage availability
Changing abundance, productivity, and structure of communities

Ecosystem

Interactions of changing disturbance and habitat size/geometry
Changing structure of communities with novel landscapes
Complex feedbacks with human adaptation such as water development with expanding agricultural growing seasons or demands linked to population growth and increasing storage needs
Counterintuitive results and surprise, trophic cascades

Direct effects. Direct effects of climate will be those that impose limits on the distribution or performance of organisms and, by their nature, should lead to the most obvious responses. Temperature, for example, can impose clear limits through lethal conditions but can also influence the performance of individuals through scope for activity and growth (Coutant 1999), embryonic development, and physiological rates that are key to life history transitions (e.g., Jonsson and Jonsson 2009). The thermal tolerances of individual organisms reflect these basic processes. Thermal preferences, or the selection of thermal environments by individuals, often reflect conditions that are optimal for growth and relative performance or success across thermal gradients. Thermal tolerances, or the physiological responses to temperature, have been quantified for many fishes through laboratory trials (e.g., Richter and Kolmes 2005; Selong

and others 2001), but they might also be inferred from observations of behavior, life history timing, and distribution along thermal gradients (e.g., Eaton and others 1995; Isaak and others 2010). Important differences exist in thermal tolerances and efficiencies within and among families, genera, and species (Crozier and others 2008; McCullough and others 2009; Selong and others 2001), although differences are generally smaller within than among taxonomic groups (McCullough and others 2009). Thermal tolerances help explain the broad patterns for species occurrences and persistence, and there are predictable patterns in species geographic ranges and longitudinal distributions within riverine networks and along thermal gradients tied to latitude and elevation (Nakano and others 1996; Paul and Post 2001; Shuter and Post 1990). In general, salmonids and some sculpins are associated with colder (and higher elevation or latitude) waters, while minnows, suckers, and sculpins are often found in warmer waters and mid to lower elevations, although all of these species can range widely through a system depending on season (Crozier and others 2008; Reeves and others 1998). Many of the introduced species that include basses, sunfishes, perches, and catfishes are commonly found in the lowest elevation and warmest waters across the western United States (Lee and others 1997).

Temperatures that commonly exceed physiological thresholds or lethal limits will presumably represent relatively hard limits to species occurrence, although variation in life history and behavior, such as the increasing exploitation of thermal refugia, may mitigate hard constraints (e.g., Crozier and others 2008; Keefer and others 2009). Organisms and populations that occur near, or more frequently encounter, their thermal limits, however, seem more likely to fair poorly in a warming world. Chronic warming should lead to

increased mortalities and shifting habitat distribution or range limits (presumably northward or upstream; e.g., Jonsson and Jonsson 2009; Rieman and others 2007; Shuter and Post 1990). Isaak and others (2010) confirmed that shifts in thermal habitats were occurring but also found that effects on species can differ dramatically within the same river network. For a species like bull trout, which is often confined to the coldest headwater streams, warming trends may cause a net loss of habitat (estimated to be 8 percent to 16 percent per decade; Isaak and others 2010) because areas are not available farther upstream to replace those that are becoming unsuitably warm (figure 16). But for species like rainbow trout, which are constrained by temperatures that are too cold in upstream areas, warming trends may merely shift habitats toward higher elevations without causing a net gain or loss in the total amount of habitat (figure 16). Recent episodes of mortality in adult salmon linked to unusually high temperatures and low flows

a) Rainbow Trout

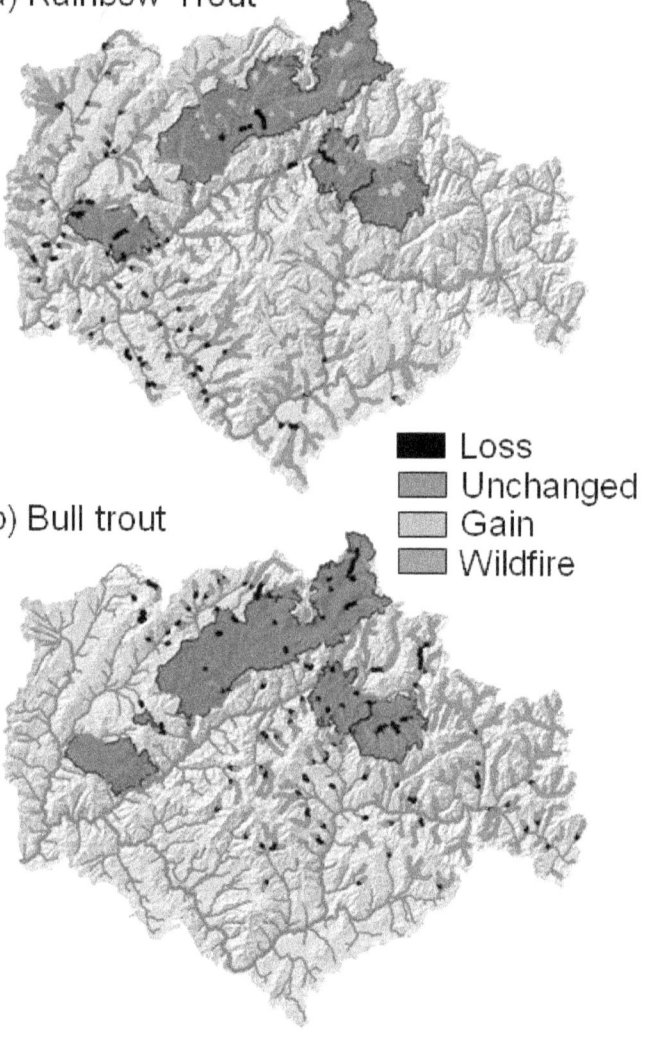

Figure 16. Shifts in the distribution of thermally suitable habitat for (a) rainbow trout and (b) bull trout spawning and early juvenile rearing in the Boise River basin from 1993 to 2006 due to long-term climatic trends. Mean summer stream temperatures greater than 11 °C and less than 14 °C were used to delineate suitable habitats for rainbow trout; whereas temperatures less than 10 °C were used for bull trout (from Isaak and others 2010).

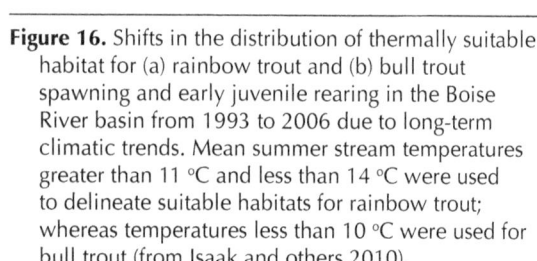

■ Loss
▨ Unchanged
□ Gain
□ Wildfire

b) Bull trout

in the Klamath (Bartholow 2005) and Columbia rivers (Coutant 1999; Keefer and others 2007), are evidence that climate change is starting to cause acute thermal stress near downstream distribution limits and be emblematic of future changes in many populations.

At finer spatial scales, thermal tolerances may be imprecise predictors of species occurrences or abundances. For example, the probability of finding juvenile bull trout often declines dramatically in streams that exceed summer mean temperatures of 10 °C to 12 °C (Dunham and others 2003; Isaak and others 2010), but individuals still sometimes occur in much warmer water. In other cases, species may persist in streams that commonly exceed their perceived thermal limits (Tiffan and others 2009; Zoellick 1999) perhaps because of increased availability of food, lack of competition with other species, or adaptations that better exploit thermal refugia or shift timing of life history transitions (Crozier and others 2008; Jonsson and Jonsson 2009). Temperature clearly has an important influence on general limits of distributions, but the complexity of interactions with other biological and physical processes are also important to consider.

Flow also may directly constrain the distribution and performance of fishes based on timing and volume of runoff, water velocities, and their effects on suitability of habitats for incubation, holding, foraging, and migration. Depending on life history stage, the frequency of extreme events associated with stream drying (Labbe and Fausch 2000) or the timing and distribution of flooding and scour events can be important (Fausch and others 2001; Jonsson and Jonsson 2009; Montgomery and others 1999).

Interactions and indirect effects. Important, but less direct, climate effects emerge through the interaction of biological and ecological processes that are influenced by temperature and flow at the scale of individual organisms, populations, habitats, and streams. For example, temperature fundamentally controls potential growth, but because the responses are nonlinear and depend on the amount of food, activity, and past conditions, warming may lead to increases or reductions in growth (Coutant 1999; Jonsson and Jonsson 2009; Richter and Kolmes 2005). Changes in growth are in turn linked to changes in age and timing of maturation, fecundity, and reproductive potential (Holtby 1988). Temperature can influence reproductive success through the rates of embryo development and timing of emergence, dispersal, and migration in relation to seasonal flooding or scour (Fausch 2008; Fausch and others 2001); the availability of food following emergence; the extent of seasonal

growth; and the timing of migrations or critical life history transitions (Brannon and others 2004; Coutant 1999; Crozier and others 2008; Jonsson and Jonsson 2009; Yarnell and others 2010).

Because the demographic growth rate of a population is ultimately an integration of reproduction and survival of individuals, stream temperatures can constrain the occurrence and distribution of populations well within their absolute physiological limits (see Beever and others [2010] for a relevant discussion regarding a small mammal), but the level of that constraint will vary. The observance of broad thermal tolerances has commonly led to predictions of local population declines and the general expectation of contracting distributions for native salmonids across large regions (Flebbe and others 2006; Jonsson and Jonsson 2009; Rahel and others 1996; Rieman and others 2007). In reality, the local responses will depend on both species and environmental context. The interaction of temperature and flow directly influence conditions in migratory corridors for both juvenile and adult salmonids, but the capacity of individuals to continue to exploit existing or new food resources or thermal refugia and for populations to compensate for increased mortality can exacerbate or ameliorate the immediate effects (Crozier and others 2010). Salmon biologists have long argued that deteriorating flow and temperature conditions have negatively influenced travel rates and survival of migrating smolts (ISAB 2007), but some populations have fared better than others in response (Connor and others 2005; Williams and others 2008). Some salmonids, such as cutthroat trout in high-elevation streams, that are commonly limited by low water temperatures or short growing seasons (Coleman and Fausch 2007; Harig and Fausch 2002) could benefit from warming. Rainbow trout and steelhead near the species' southern limit in California appear to be highly vulnerable to climate change (McCarthy and others 2009) while rainbow trout in central Idaho may be influenced relatively little (Isaak and others 2010). Highly productive populations or those that can find, gain access to, and exploit alternative habitats or refugia to seasonal climate constraints might also fare better than anticipated.

Community and ecosystem effects. Climate change will also influence native and introduced fishes through more complex interactions at population, community, and ecosystem levels of organization. Interactions among species and between species and changing landscapes, and even adaptation within species and populations, could become important (Crozier and others 2008; Williams and Jackson 2008; Yarnell and others 2010). Some changes in growth, productivity, abundance, and distribution associated with warming are likely to occur with many of the organisms in any community. Because those changes will vary among species, relative abundances and even the composition of communities in streams may shift (e.g., Hilborn and others 2003). The availability of species serving as forage and the numbers, consumption rates, and activity levels of those acting as predators or competitors may change as well. Petersen and Kitchell (2001) used bio-energetic models to show that warming of 1 °C to 2 °C in the Columbia River could result in a 26 to 96 percent increase in consumption of juvenile salmon by northern pikeminnow. Others suggest that longer growing seasons and warmer waters will allow the effective expansion in distribution, abundance, and influence of other potential competitors, disease, and predators (Hari and others 2006; ISAB 2007). Warming in the Columbia River has been associated with a rapid expansion in populations of American shad, for example, that are anticipated to consume zooplankton prey, which are important for growth and survival of migrating juvenile fall Chinook. Northern pikeminnow that may also act as important competitors with juvenile salmonids in smaller tributary streams might be expanding further into upstream salmonid rearing areas (Reese and Harvey 2002). Introduced parasites and diseases may expand with warming temperatures and could play an increasingly important role as additional constraints on the distribution of some species (Franco and Budy 2004; Hari and others 2006).

The interaction of temperature with flow and other physical processes may lead to new patterns of disturbance that will influence the resilience and persistence of broader habitat and population networks. Changing flow regimes, more extreme weather events, and increasing frequency of large wildfires suggest important changes in the extent, frequency, and magnitude of disturbance for aquatic environments (McKenzie and others 2004; Isaak and others 2010). The aquatic ecosystems associated with streams and rivers of the Rocky Mountains evolved in highly dynamic landscapes. Wildfire, flood, drought, mass erosion, glaciation, and volcanism have been important in shaping these systems over ecological and evolutionary time scales (Bisson and others 2009; Kirchner and others 2001; McPhail and Lindsay 1986; Waples and others 2008). In an ecological and evolutionary sense, disturbance and even radical changes in climate are nothing new. But in many cases, the fundamental mechanisms that are contributing to the resilience and adaptive capacity of native populations (i.e., connectivity among diverse

habitats and populations, local adaptations, and the broad expression of genetic and phenotypic diversity) have been eroded or lost through the myriad effects of human development, habitat loss and fragmentation, and the introduction and continuing invasions of new species (Bisson and others 2009; Rieman and Dunham 2000; McClure and others 2008). For some species and populations, climate change could exacerbate fragmentation (Fagan 2002; Rieman and others 2007), leading to smaller and more isolated habitats and, simultaneously, to larger more frequent disturbances (Isaak and others 2010). In fact, Jentsch and others (2007) argue that uncharacteristic disturbance will play the dominant role in structuring ecosystem response to climate change as some species and populations that once flourished in a complex and highly dynamic world may no longer have the necessary resilience.

Biological adaptation may help. Most of the efforts to consider the implications of climate change assume that species and populations will continue to use and respond to the environment as they have in the recent past. In some instances, biological adaptation to changing environments could mitigate some of the challenges organisms face. Change in landscapes, watersheds, and ecological systems that support native fishes is nothing new. In some cases, these changes have been rapid and dramatic such as those linked to large wildfires. In others, they have been more gradual as climates have shifted in the past. Large-scale dispersal from post-glacial refugia (McPhail and Lindsey 1986) and evolution through natural selection have played central roles in the adaptation of native fishes and communities to changing environments and climate in the past (Waples and others 2008). The pace of change and the capacity for resilience in many populations is arguably much different than it was through most of the natural history of the species and communities we manage today, so the capacity for rapid adaptation has become an important consideration of ongoing research (e.g., Williams and others 2008). There is growing evidence that many fishes can adapt relatively quick to changing conditions through behavioral or phenotypic plasticity and rapid evolution (Crozier and others 2008), although evolutionary scope varies among different traits (McCullough and others 2009).

Many salmonids can colonize or exploit new habitats almost as they become available (Isaak and others 2006, 2007; Milner and others 1987, 2000, 2008). Species like bull trout that exhibit extensive movements linked to spawning, foraging, and overwintering appear to be flexible in their use of distinct environments (Brenkman and others 2007; Volk and others 2010). If recent patterns of ranging or foraging are no longer profitable, they may have the capacity to shift patterns of movement until they find ones that are. Chinook salmon have been observed to use new spawning habitats created by debris flows that are not in immediate proximity to traditional spawning sites (Isaak and Thurow 2006; R. Thurow, personal communication of unpublished data). Similarly, bull trout that are blocked from past spawning streams can exploit new streams (Neraas and Spruell 2001). Even though climate change may reorganize the distribution and availability of critical habitats, some species and populations may be able to effectively exploit what emerges, as long as suitable habitats still exist and are accessible.

Fishes also have some capacity to adapt to change in place. Often, multiple life histories exist within a population or among several closely allied populations. Some may be favored by new conditions while others are not. Hilborn and others (2003) summarized long-term variation in relative success of three distinctly different life histories for sockeye salmon in a large lake system in response to climate variability. The expression of diversity in life history stabilized overall sockeye production even as individual stocks responded to changing conditions. Plasticity or flexibility in life history can also emerge within individual populations over short time scales (e.g., within a single generation). Because growth and other population responses are linked, changing growth rates can influence population dynamics and resilience. For example, rainbow trout growth and maturation rates increase substantially in warmer post-fire streams, which is associated with earlier maturation and could increase population growth rates and recovery in response to an initial disturbance (J. Dunham, personal communication of unpublished data). Heck (2007) and Kennedy and others (2003) observed similar responses in other salmonids. Distinct life histories could be lost with changing climates (Beechie and others 2006), but if new environments provide conditions that are within the range of possibilities for a species, the phenotypic plasticity that allows populations to adapt quickly may still exist even if it has not been important in the recent past (e.g., Healey and Prince 1995).

Finally, evolution through natural selection could be important as well. There is concern that rates of climate change will overwhelm capacity for evolution through natural selection in remnant populations (e.g., Beever and others 2010; Crozier and others 2008). But there is also growing evidence that important evolution can occur in fishes within 10 to 20 generations (40 to 80 years for species with a 4-year generation time; Hendry and others 2000; Stockwell and others 2003; Waples and others 2007). Changes in spawning and migration timing and the expression of distinctive life histories required to exploit different environments may be particularly responsive to natural selection (Crozier and others 2008; Quinn 2005). Recent work suggests that fall Chinook salmon in the Snake River, for example, are evolving novel rearing and migration timing in response to changes in flow and temperature caused by water development over the last 40 years (Williams and others 2008). There is also some evidence that local adaptation may be possible in thermal tolerances or growth efficiencies for some salmonids, but the capacity for rapid evolution in these traits is unclear and may be more limited than with others (McCullough and others 2009).

Summary

The bottom line is that change is happening and will continue, though it will not always be intuitive or easily predicted. For many fish species, suitable habitats defined by temperature, flow, and other physical or biotic conditions may simply shift in location (e.g., to higher elevations or latitudes). Some species, populations, and communities may be able to track these changes and simply "relocate," but barriers to dispersal and migration will limit many others. Interactions between biological and physical process may also lead to results that are difficult to anticipate, even for well-studied species. Jager and others (1999), for example, used individual-based bioenergetic and population models to show that changes in foraging, growth, and scour could lead to counterintuitive results in the distributions of co-occurring brown trout and rainbow trout in streams of the Sierra-Nevada mountains. In other cases, the interaction of climate change, heterogenous landscape responses, shifting species distributions, and the new suite of potential biotic and physical interactions will lead to novel environments, trophic cascades, and communities with no natural precedent and little foundation for prediction (Williams and Jackson 2008; Williams and others 2007). Adaptation to changing environments through natural selection and plasticity is possible, but the speed and capacities for adaptation are not well known and may be outpaced by the rate of climate-driven environmental change.

What Can We Do About It?

Already, there is considerable literature that offers guidance on how natural resource managers might respond to climate change (e.g., Furniss and others 2010; Hodgson and others 2009; Millar and others 2007; Nelitz and others 2009; Noss 2001; Welch 2005). Some documents are focused on particular ecosystem issues such as coral reef bleaching (West and Salm 2003), reorganization in montane forests (Millar and others 2007), and salmon conservation (Bisson 2008; ISAB 2007; Schindler and others 2008). But common concepts and elements of conservation and restoration management emerge repeatedly in much of the discussion (Mawdsley and others 2009).

Two common themes in management response to climate change are "adaptation" and "mitigation" (Millar and others 2007; USDA Forest Service 2008). The *Forest Service Strategic Framework for Responding to Climate Change* (USDA Forest Service 2008) defined adaptation as "actions to adjust to and reduce negative impacts of climate change on ecological, economic and social systems." The *Framework* discussion considered actions that may support conservation of existing ecological elements as well as transition to fundamentally new conditions. We follow others in distinguishing facilitation as action to move the system to a new state and adaptation as action to conserve an existing state. In this context, we define adaptation as management to help populations, species, communities, and ecosystems absorb climate change with limited alteration of the structures, functions, or services we value. The terminology can be confusing, but from the perspective of native fish conservation, we suggest that adaptation is essentially an effort to conserve species, populations, communities, or fisheries as, and where, they currently exist. We might focus, for example, on actions that would help ensure persistence of a particular species, population, or fishery by enhancing resistance and resilience to changing climate and its effects.

Mitigation in this context reflects the efforts that governments, agencies, and the public make to reverse or slow the causes of climate change such as energy conservation, greenhouse gas reductions, and carbon sequestration. Although land and forest managers are heavily engaged in mitigation efforts (e.g., wildfire management, reforestation, and stand management

for carbon sequestration) that can directly influence aquatic ecosystems, this synthesis focuses primarily on adaptation strategies for aquatic ecosystems. Rieman and others (2010) consider some issues in coordinated management of terrestrial and aquatic systems that may emerge in response to management that is focused on mitigation.

In the remainder of this section, we outline the general guidance for adaptation and facilitation in response to climate change in aquatic systems in five topic areas: (1) enhance resistance and resilience; (2) prioritize limited management resources; (3) facilitate transitions to new populations or communities when appropriate; (4) develop local information; and (5) coordinate management among entities, resources, and actions (table 2). We consider each of these in turn.

Enhance Resistance and Resilience

Resistance and resilience are ecological concepts that reflect the capacity of natural systems to absorb or recover from environmental change or disturbance and thus to persist into the future. There is considerable literature devoted to the characteristics of resistant and resilient biological systems (e.g., Harrison 1979; Holling and Meffe 1996) and management efforts to conserve or restore those characteristics in the face of consumptive exploitation (e.g., Healey 2009), habitat loss and disruption (e.g., Bisson and others 2009), and changing environments (e.g., Healey and Prince 1995; Hodgson and others 2009). The terms are often linked and are sometimes used interchangeably but are, nevertheless, the essence of managed adaptation to climate change.

In our perspective, resistance represents the capacity of important habitats, populations, or communities to absorb an environmental shift or disturbance with limited or negligible deflection in abundance, structure, or function (figure 17; e.g., West and Salm 2003). For example, some streams may show limited response to increasing air temperatures because they are well buffered by the influence of groundwater or snowmelt (Boxall and others 2008; Mohseni and others 1999). Alternatively, some populations may not change in response to streams that warm substantially

Table 2. Examples of management options to support adaptation of salmonid fish populations and stream communities to the effects of climate change (see also Bisson 2008; Furniss and others 2010).

Management option	Rationale or anticipated effect
Enhance resilience and resistance	
Reduce non-climate stresses	
Maintain or restore instream flows and natural hydrologic regimes	Maximize available habitat, increases terrestrial interactions, and buffers streams against temperature increases and the loss of habitats during low flow events.
Maintain forest and vegetative cover to reduce rain-on-snow flooding and delay snow melt	Mitigate loss of snow pack storage, earlier runoff, and reduced summer low flows. Conserves forest, wetland, and riparian areas that tend to store water for later summer base flows.
Maintain or restore riparian, floodplain, and wetland conditions and connections with streams; reintroduce beaver	Maximize stream shading, bank stability, terrestrial food inputs, and recruitment of woody debris that helps form diverse habitat; enhance water storage for delayed summer discharge during warm, low flow periods.
Protect and restore critical or unique habitats that buffer survival during vulnerable periods seasonally or in the life history	Ensure that nodes connecting seasonal or complimentary habitats or refugia do not become bottlenecks to production. Off-channel habitats, spring brooks, and seeps important as early rearing environments; flood or thermal refugia and stream segments important as connections; broader expanses of habitat are examples.
Disconnect roads from the drainage network, and remove roads and dikes that constrain or disconnect channels and flood plains	Buffer the effects of peak flow events.
Limit or stop introduction and expansion of non-native species	Reduce potential competitors, predators, diseases, and hybridization that may constrain habitat capacity, individual growth rates, and survival.
Eliminate or control pollutants or contaminants	Reduce stresses associated with eutrophication, toxic materials, or other effects on growth, productivity, and survival.
Conserve and expand the size of habitat networks and migratory connections	
Remove or modify barriers to fish movements	Allow individuals to move freely and track suitable habitat distributions or re-colonize disturbed areas. Allow full expression of alternative life histories and increased productivity of migratory forms.
Maintain or reconnect large networks of habitat	Larger habitats support larger populations that are less susceptible to extreme events and loss of genetic diversity; they are also more diverse and capable of supporting other needs outlined above and below.
Conserve genotypic/phenotypic diversity	
Conserve or restore a diverse representation of habitats across river basins	Provide biological resilience and increase odds that some populations or individuals will be adapted to future conditions or have the capacity to evolve.
Conserve or restore large networks (see above)	Maintain large population sizes to minimize loss of genetic variability and adaptive potential.
Prioritize	
Clarify goals and values	Minimize confusion and conflict among disciplines and agencies, increasing the chances of recognizing conflict and opportunity before the fact.
Focus on populations as units of conservation	Ensure that some populations are as resilient or complete as possible rather than moving all incrementally and leaving all vulnerable to extreme events.
Weigh vulnerability	Focus on the greatest benefits for the least cost; avoid lost causes.
Consider relative values and balance resilience, representation, and redundancy	Ensure maintenance of some populations and as much diversity as possible in the face of extensive change or large, catastrophic events.
Take no action	Conserve limited conservation resources so they can be expended in the most beneficial areas.

Table 2. *Continued.*

Management option	Rationale or anticipated effect
Facilitate transition to new states	
Human-assisted migration	Transport individuals to existing but otherwise inaccessible habitats or refugia to maintain gene flow, establish or re-establish self-sustaining populations, and buffer potential for catastrophic losses.
Remove barriers to invasion	Where native species can clearly no longer persist, allow colonization by new species that may be better suited to new environments and still provide some ecological function and value.
Introduce new species	See "Remove barriers to invasion."
Develop local information	
Understand context	Recognize regional and local trends in climate that provide a context for the past and future changes that are relevant to the local systems of interest.
Synthesize existing information	Understand changes or trends in habitats or populations that have already occurred in local or nearby representative systems. Indentify important gaps in information for further work.
Model to fill gaps	Extrapolate likely or potential changes based on current models. Use the models as a basis for thinking about how change may occur but recognize limitations and uncertainty.
Monitor and document trends	Strengthen local knowledge; test, validate, or reject models, predictions, and hypotheses. Review, revise, and refine management.
Coordinate efforts	
Across disciplines and resource values	Recognize potential conflicts and opportunities to focus high-resolution analysis where actually needed; mitigate activities and recognize opportunities to leverage common values and supporting work.
Across agencies and jurisdictions	Maximize environmental range of suitable habitats used by native species and decrease chances of hybridization in some instances.
With the public	Build understanding of ecological values in aquatic systems and support for the actions and tradeoffs that inevitably must be addressed.

Figure 17. Resistance and resilience reflect the capacity of a population to absorb and recover from disturbance.

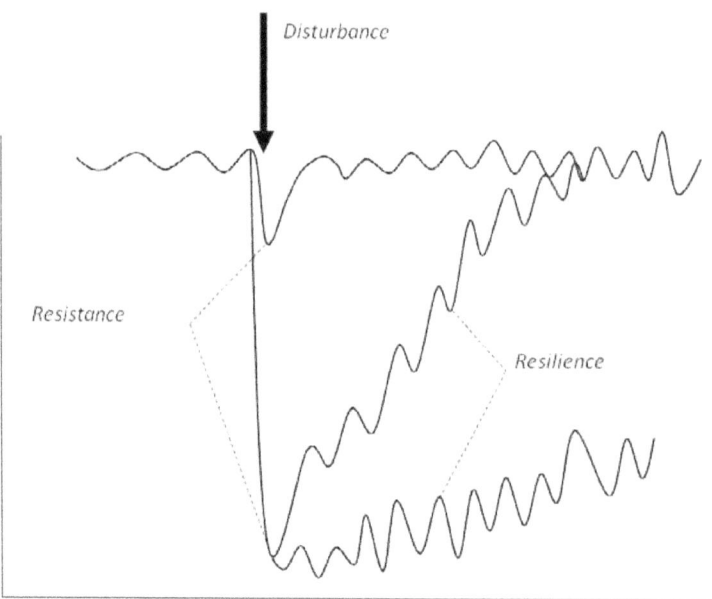

if temperatures remain within the range of thermal tolerances. Some fish populations may show limited response in overall number to increased mortality caused by environmental change that is substantial because of strong density-dependent compensation in juvenile recruitment (Crozier and others 2010).

Resilience can be viewed as capacity for, or rate of, biological and ecological recovery from change or disturbance that causes substantial reduction in abundance. Resilient populations, for example, retain the capacity to recover from overexploitation or short-term disturbances such as wildfires or floods (e.g., Dunham and others 2003; Rieman and Clayton 1997) even though population size or structure might have been depressed to only a fraction of pre-disturbance levels. Resilience might also be reflected in the capacity of populations to adapt to longer-term changes in environments by altering behavior, life history, or other characteristics in ways that allow them to recover and persist even though conditions remain fundamentally different than in the past. As suggested earlier, biological adaptation that contributes to the resilience of a population or species might occur either through behavioral or phenotypic plasticity or through natural selection and rapid evolution (Healey 2009; Waples and others 2009).

Much has been written about the conditions and characteristics of resistant and resilient systems, particularly in the contexts of conserving and restoring commercial fisheries (e.g., Healey 2009), conserving biological diversity in response to managed landscapes (e.g., Bisson and others 2009; Duffy 2009; Holling and Meffe 1996), changing fire regimes (e.g., Bisson and others 2003; Dunham and others 2003; Gresswell 1999; Rieman and Clayton 1997), and changing climate (Noss 2001). Based on that literature, adaptation to restore or conserve resistant and resilient native fish populations might be distilled to four key areas: reduce non-climate stresses, conserve and expand critical habitats, reconnect streams and habitats, and conserve genetic and phenotypic diversity. These are considered below.

Reduce non-climate stresses. Decreasing traditional stresses of aquatic systems that contribute to reduced growth or survival of individuals or reduced capacity of habitats can lead to increased potential growth rates in local populations and increased resistance and resilience to climate change effects (Wooldridge and Done 2009). These are actions that fisheries biologists have focused on for decades. Populations that are productive and have the capacity to absorb fishing are likely to be resilient and resistant to changing environments that impose additional mortality or restrict habitat

capacity. Efforts to conserve watersheds and habitats that already are productive and highly functional are likely to be the most effective and efficient steps, but restoration of watershed processes that support the creation and maintenance of complex and productive habitats can be important as well (Beechie and others 2010). Habitat degradation has been a central problem in conservation and management of native fishes and fisheries, so opportunities to restore more productive environments are often widespread. Efforts to restore stream flows (Van Kirk and Benjamin 2001); remove or mitigate contaminants, pollutants, and other extraneous sources of mortality (e.g., Peterson and others 2010); reconnect streams with their floodplains; and restore riparian functions (especially those important to flow and temperature) are all examples of important work in the region. Rahel and others (2008) suggest that efforts to control invasive species that may act as predators and competitors is also an important tool, and substantial interest has focused on this issue in recent years (e.g., Eby and others 2006; Fausch and others 2009; Muhlfeld 2009; Peterson and others 2008; Shepard 2002). Reducing the threats of invasion through direct reductions of source populations or through construction of barriers to dispersal may be more effective than control of widely established populations (Meyer and others 2006). There are, however, important tradeoffs to consider with use of barriers designed to limit the expansion of non-native species because these will also limit dispersal or expansion of native species (see "Reconnect streams" below).

Conserve and expand critical habitat. Larger habitat areas or capacities will help expand the size and diversity of populations and the diversity of communities (Hodgson and others 2009; Neville and others 2009). A growing body of work shows that salmonids are more likely to persist indefinitely in larger or more complex habitat networks (Fausch and others 2006). Larger networks are more likely to provide complementary habitats required to complete life histories; internal complexity and area needed to absorb catastrophic disturbances without loss of the entire population; and greater genetic and phenotypic diversity that can facilitate adaptation (Fausch and others 2009; Hodgson and others 2009; Neville and others 2009). In some cases, climate change may result in the expansion of suitable habitats, but for many coldwater and headwater species and populations, changes will result in no net change or further shrinking of what are already small, highly fragmented habitat networks (Fausch and others 2006; Isaak and others 2010; Rieman and others 2007). For example, many native salmonids in Rocky Mountain streams are already constrained to colder headwater reaches and are commonly found in remnant stream networks that consist of a few kilometers at most (e.g., Brown and others 2001; Fausch and others 2006; Harig and Fausch 2002). For a species such as bull trout, which appears to be constrained at its upstream extent by stream gradient or size, many populations are essentially against the wall (Isaak and others 2010; Rieman and others 2007). Species that occupy shrinking habitats sensitive to warming or more extensive dewatering could change even more quickly than anticipated if disturbances linked to fire, large storms, and drought become more frequent or extensive (Brown and others 2001; Isaak and others 2010; Jentsch and others 2007).

Conserving the size and extent of existing, high-quality habitats and habitat networks will be an important step wherever possible. But expanding those by removing constraints or restoring processes that create or maintain productive habitats adjacent to good condition areas could be important to gain areas large enough to absorb the effects of a changing climate. Important questions still exist about how much area is enough to conserve productive, resilient populations. Maintenance of genetic diversity and adaptive potential will require enough habitat to support hundreds of adults at a minimum (Rieman and Allendorf 2001), but persistence in varying environments may require considerably more. Some salmonids have persisted in habitats that are limited to a few kilometers of stream (Hastings 2005; Morita and Yamamoto 2002), while others may need an order of magnitude more (e.g., Dunham and others 2002; Dunham and Rieman 1999; Fausch and others 2006). There are few rules of thumb to guide decision making for all the species and communities important across the Rocky Mountains, but conserving and creating habitat networks larger than a few kilometers wherever and whenever possible will likely be a fundamental step in climate adaptation (e.g., Hodgson and others 2009).

Reconnect streams. Removal of barriers to fish movement can be an important step in the expansion of habitat networks, but it will also help re-establish the full expression of migratory life histories and connections among populations. Many fishes native to the Rocky Mountain West, particularly the salmonids, express a variety of life history strategies that include a broad range of movements among complementary habitats (Brenkman and others 2007; Northcote 1997; Rieman and Dunham 2000). Juvenile migration to productive rearing areas and adult migration and homing to natal areas for anadromous, fluvial, and adfluvial life histories have been widely documented and explored in the salmonid literature for decades (e.g., Quinn 2005). Expression of migratory behavior can contribute to resilience of populations through larger adult body size, fecundity, recruitment, and potential population growth rates (Peterson and others 2007; Rieman and Dunham 2000). It can also represent a diversity of strategies that effectively hedge against disturbance (e.g., Rieman and Clayton 1997) and environmental variability (Hilborn and others 2003) likely to accompany climate change. Connection and migratory life histories are also key to linkage among populations (Dunham and Rieman 1999). Metapopulation theory has received substantial attention in the ecological literature over the last two decades, and there is growing evidence that movement among populations, most likely through juvenile dispersal or straying of migrant adults, is important to gene flow, maintenance of genetic diversity, demographic support, and recolonization of disturbance-prone or recently available habitats (Dunham and Rieman 1999; Fausch and others 2006; Isaak and others 2007; Letcher and others 2007; Neville and others 2006, 2009).

There are concerns that removal of barriers will facilitate invasions of non-native species (Fausch and others 2006, 2009). Non-native species are an important threat to persistence of many native fishes throughout the Rocky Mountain West (Eby and others 2003; Fausch and others 2006, 2009; Rahel and others 2008), and the intentional use of migration barriers may be an important strategy against non-native invasions that could be accelerated with changing climate (Jackson and Pringle 2010; Rahel and others 2008). The tradeoffs with isolation, however, are clear as well and must be carefully considered by managers who are contemplating intentional isolation (Fausch and others 2006; Peterson and others 2008). If isolation is a serious option, the area and quality of the habitat to be isolated and the potential influences of climate change on those characteristics will be important constraints to consider. Peterson and others (2008) provide one framework for assessing the tradeoffs.

Conserve genetic and phenotypic diversity. Maintaining biodiversity is important to ensure the greatest possible capacity for natural biological adaptation to variable and changing environments. As we discussed earlier, adaptation can occur through varied success of distinct life history strategies, short-term plasticity in phenotypic or behavioral characteristics, and natural selection and evolution. Regardless of the mechanism, conservation and restoration of biological diversity are fundamentally important strategies for managers faced with climate change and other natural threats (Hodgson and others 2009; Levin and Lubchenco 2008; McClure and others 2008). Because the precise nature of change and the conditions that favor different life histories are largely unknown, conservation or representation of adaptive potential and the fullest range of genetic and phenotypic diversity possible is prudent. Maintenance of genetic diversity will depend largely on the size of local populations and potential for gene flow among populations (e.g., Neville and others 2009; Rieman and Allendorf 2001), so the extent and connection of habitats considered above will be important here as well. Genetic and phenotypic diversity will also depend on representation of populations (and their habitats) across diverse environments that reflect as much of the full genetic variation, local adaptation, and differential phenotypic expression as possible (Allendorf and others 1997; McClure and others 2008; Healey 2009). If details of population biology are not available, representation of populations across distinct gradients of productivity and growth (see McGrath and others 2008), hydrologic regime (Beechie and others 2006), disturbance history (Waples and others 2009), or spawning habitat (Hilborn and others 2003) could be useful starting points.

Prioritize Limited Management Resources

One of the most important problems for conservation and restoration in the context of climate change will be the allocation and effective prioritization of limited management resources. In general, the list of known or anticipated habitat and watershed problems is far larger than available funding or logistical or technical capacity can address. This is a widely acknowledged problem for conservation biology, where the concepts of triage and prioritization of limited resources have been a focus of research, management application, and some debate (e.g., Allendorf and others 1997; Bottrill and others 2009). It is a challenging problem for fisheries and aquatic managers in the West as well. The prioritization of some habitats, populations, or streams over others can be difficult because it may imply giving up on some to focus on others. Focusing financial resources rather than giving everyone a piece of the pie can also be subverted by political or professional provincialism. The alternative of spreading limited resources across as many habitats or populations as possible, however, may only guarantee that all remain compromised and vulnerable to the challenges of climate and other environmental change in the future (Bottrill and others 2009; Frissell and others 1997; Reeves and others 1995; Rieman and McIntyre 1995).

There is an extensive literature in conservation biology and restoration ecology considering prioritization (e.g., Fausch and others 2006; Groves 2003) and a growing effort to place it in the context of climate change (Nelitz and others 2009; Noss 2001). A common element in many discussions has been the concept of risk (Fasuch and others 2007; Francis and Shotton 1997) or vulnerability, which can be defined as the probability that something of significant value will be lost as a result of a potential change. Clarifying risk or vulnerability can help identify priorities, but requires some sense of the relative value in different watersheds, habitats, species, populations, or other targets of conservation or restoration management and the sensitivity or probability that they will be lost if change occurs (see the following Text Box).

Considering Vulnerability

Terms like *vulnerability* and *risk* have received considerable attention in the climate-adaptation literature. In the context of this report, we consider them to be essentially equivalent. Formally, risk has been considered the product of two conditions: the *value* placed on a resource or condition of interest (i.e., the consequences of climate change) and the *probability* of losing that resource if an event or change occurs (in this case, the effects of a changing climate; e.g., Wood and others 2008). Vulnerability has been defined as the product of *sensitivity* and *exposure* (e.g., Furniss and others 2010), where sensitivity reflects both value and the probability of change or departure in that value given change in climate, and exposure is the expectation or probability that a change of a given magnitude will occur. In this sense, risk and vulnerability might be used as shorthand for the chances of losing or substantially altering specific, important habitats or populations. If the probability for change, sensitivity to that change, **or** the value in question is low, risk or vulnerability will be low. If the probability, sensitivity, **and** value are high, vulnerability or risk will be high. Presumably, some knowledge of the natural resource values most sensitive to climate change can help managers prioritize limited funding and other management resources (see section on Prioritizing Limited Resources). Structured decision analyses or formal risk analyses provide quantitative frameworks to consider important tradeoffs. Actions that are anticipated to produce the largest benefit (reduction in vulnerability or risk) for the least cost (e.g., high benefit/cost relative to other alternatives) may be logical priorities.

Formal decision models are becoming more common in the natural resource literature with some examples now in regional aquatic or fish and wildlife issues (Marcot and others 2001; Peterson and others 2008; Reckhow 1999). Models that quantify probabilities and the relationships among driving and dependent variables can require detailed information that may be unavailable or difficult or expensive to collect. A variety of approaches have been developed, however, that allow incorporation of data and established relationships as well as professional judgment (Peterson and others 2008). Whether decision models are better served by hard information or general experience can be debated, but an important point remains—any management decision is based on some model of how the world (or habitats and populations) works. That model may be a highly complex mathematical representation, a simpler argument of logic (e.g., if this, then this), or

something in between. Regardless of the approach, all models start with some effort to articulate the basic logic and important relationships. That effort alone can be important as biologists and managers attempt to communicate their rationale, weigh tradeoffs, and identify critical uncertainties for further work. *Influence diagrams* are a useful tool in this process and can become a foundation for more quantitative models or discussion and refinement of logic that guides management actions.

In the figure below, we present an influence diagram that represents a "model" of persistence of a local fish population in response to climate. This model reflects our experience with native salmonids like cutthroat trout and bull trout in the Rocky Mountain West. We have outlined the interactions of natural and human-caused disturbances with habitat and population characteristics that we believe influence the resilience of populations in streams across the region (e.g., Dunham and others 2003; Dunham and Rieman 1999; Fausch and others 2006; Isaak and others 2007, 2010; Peterson and others 2008; Rieman and Dunham 2000; Rieman and others 2003, 2007). The influence diagram focuses on *population persistence* or the probability that a local population of native salmonids will persist for some extended period given a set of environmental conditions and a particular climate. In our view, population persistence will be determined by essentially three conditions:

- *Population growth rate*—reflects the capacity of the local population to absorb environmental stresses and resist or rebound following disturbance.
- *Adjacent populations*—reflects the connectivity to surrounding populations that can be sources of gene flow and recolonization or rescue if the immediate population cannot maintain itself.
- *Exposure to a catastrophic event*—reflects the chance that the population can be reduced to a very low level by a natural or human-caused disturbance, which increases as the population is restricted to a smaller area or size or as disturbances become larger or more frequent in relation to that area or size.

The conditions represented by each of these nodes will in turn be influenced by the conditions in the environment that define the extent and quality of available habitats; interaction with other species; and the frequency, magnitude, and extent of watershed disturbances that could threaten the persistence of a population confined to a limited area. By highlighting the nodes or conditions that can be directly influenced

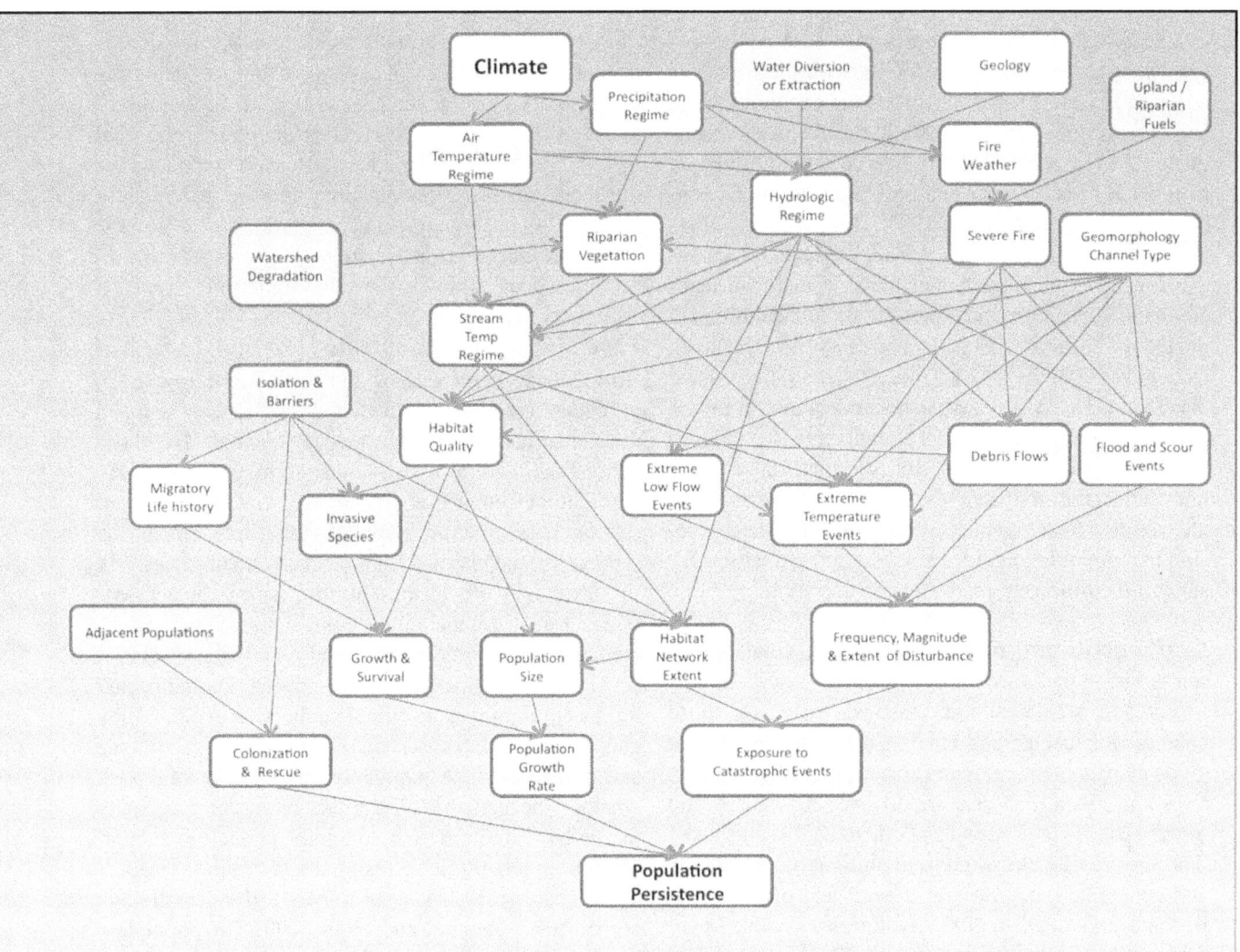

An influence diagram showing linkages among climate, environmental conditions, habitat, and the anticipated persistence of a local population of native salmonids. Vulnerability of a local population would be represented by the probability of persistence and could be combined with a measure of the "value" of that population to represent risk. The shaded nodes represent conditions that could be influenced by management and thus opportunities for action that support adaptation to a changing climate.

by management actions, it's possible to see the underlying logic and assumptions that ultimately justify any adaptive actions. For example, removal of a migration barrier could lead to expansion or reconnection of the population with surrounding habitats and the re-expression of migratory life histories that could offset the effects of a changing climate on habitat quality or catastrophic disturbances.

In this example, we assume that the population is valuable for conservation purposes. Population persistence, however, could be modified by a utility function or measure of the value of the population to represent a measure of risk that could be used to compare across populations or management actions. A model like this, if fully specified through a series of underlying quantitative relationships implied by the arrows, could support a risk analysis or formal decision process where the user can modify inputs that reflect changes in climate or management and evaluate the relative change in risk that results (see Peterson and others 2008 for an example). We don't suggest that this model is the correct or only one, but rather that it is the logical framework underpinning much of the discussion about the alternatives for adaptation to climate change in this report. We suggest this as a starting point to stimulate thinking about the conditions and processes important to persistence of populations that biologists and managers work with across the region.

Clarifying risk also requires some sense of the feasibility of reducing that probability with the resources at hand. Issues of scale (how much is enough; Fausch and others 2006; Peterson and others 2008; Rieman and others 2007), spatial structure (where should efforts be in relation to each other and how they should be connected; Frissell and others 1997; Hodgson and others 2009; Rieman and Dunham 2000), critical processes that constrain habitats or populations (what do you actually restore; Beechie and others 2010), and decisions in the face of uncertainty (Lawler and others 2009; Ludwig and others 1993; Pielke 2009) are also part of the broader discussion. There is no simple generalization for prioritization that lends itself to all situations (e.g., Hodgson and others 2009); and any strategy must reflect local knowledge, constraints, and conditions. However, there are common themes and concepts in much of the discussion that might help guide managers and biologists. We suggest the following points for consideration.

Clarify goals and associated values. Goals represent a vision of future natural resource conditions that reflect values society and managers place on those conditions (Beechie and others 2008). The values that guide management goals and objectives for aquatic systems and inland fisheries are diverse, but Fausch and others (2006) and Rieman and others (2010) suggested that they can be classified as evolutionary, ecological, and economic (see also Angermeier and others 1993; Beechie and others 2008). Evolutionary values may focus on evolutionary legacies and native biological diversity such as that represented by genetically pure populations of native cutthroat trout. Ecological values may focus on maintenance of ecological function and ecosystem services and capacities for resilience and adaptation to environmental change. Economic values may focus on economic return linked to tourism or sport fishing opportunity. These values clearly are not independent, but in managed systems, they are also not always simultaneously possible (Fausch and others 2006). For example, conservation of genetically pure cutthroat trout may require intentional isolation of a population threatened by invading brook trout or hybridizing rainbow trout (Fausch and others 2006; Rahel and others 2008). Alternatively, a focus on ecological function or economic contribution might view a hybridized population or a brook trout population that retains some resilience to disturbance or a fishery by virtue of larger network connections and the presence of migratory life histories as an acceptable result (see Fausch and others 2006, 2009; Peterson and others 2008; Rieman and others 2010 for extended discussion

of these issues). The point is not that one set of values is primary, but that different objectives that are seemingly allied under more general goals (e.g., conservation of aquatic ecosystems) can be in conflict (e.g., intentional isolation or transition to a new species better suited for the local environment) and may require different responses to conservation and restoration in the context of climate change. It is important to be clear about conservation goals and why they are important.

Populations as fundamental units of conservation. Management actions and restoration projects, in particular, are often relatively limited in extent (e.g., an individual habitat, stream, or road segment). But those actions will be biologically meaningful only to the extent that they contribute to increased size, resilience, or resistance of an entire population that may depend on several kilometers of stream or even an entire network of streams. A life history perspective might consider whether the habitats influenced by management actions are likely to be limiting to a local population where individuals must link spawning, rearing, foraging, and refuge environments to complete their life cycles (Bilby and others 2003; Lake and others 2007; Northcote and others 1997). A population perspective might ask whether the network of habitats represent a capacity that is large and internally complex enough to support populations that will be resilient and retain the capacity to adapt (e.g., Isaak and others 2010; Neville and others 2009; Rieman and others 2007). A population perspective also suggests that, to the extent possible, logical priorities might focus first on securing one network or population before moving on to others (Frissell and others 1997). Linked population and habitat models (Fullerton and others 2009; Honea and others 2009; Jorgensen and others 2009) are being explored as tools for prioritizing habitat conservation or restoration in a population and full life history context. The important point, however, is that to be effective, conservation and restoration will need to provide enough of the right habitat for complete populations to persist and hopefully flourish in the face of climate change. Essentially, it may be better to secure one or a few populations completely than to do a little work in many (Levin and Lubchenco 2008).

Consider relative vulnerability among populations and habitats. An important point from monitoring and model predictions is that some habitats and populations will be more vulnerable to climate change than others. Populations that occur on the margins of suitable habitat or in restricted habitat networks, for example, are probably more vulnerable than those in large networks that are well within suitable conditions or on the leading edge of habitats likely to improve in suitability. Streams that are buffered by ground water or hyporheic exchange may warm more slowly than those that are not. Streams now in transitional (rain-snow) hydrology are more likely to see a dramatic change in winter flooding than those that are well within current rain- or snow-dominated systems (Beechie and others 2006; Hamlet and Lettenmeier 2007). There is considerable effort to refine climate-hydrologic-temperature predictions to help managers consider the variability in sensitivity to climate change (e.g., Isaak and others 2010; Mantua and others 2009; Nelitz and others 2009; Wenger and others 2010). But even without those models, managers and biologists often have a good sense of the sensitivity of their respective streams and populations. Systems that have fared well in unusually warm-dry years or that remain productive despite the climate challenges of the last decade may be particularly important to secure and expand as future cores most resistant to future change. Logical priorities are systems that are least vulnerable or where vulnerability can easily be reduced (e.g., removal of migration barriers) because they represent the best chances with the least investment.

Consider relative value among conservation units.

Deciding where to focus first has been a central topic in conservation biology that depends on values and the conditions that support resilience. A common shorthand includes the concepts of resilience, representation, and redundancy (Fausch and others 2006; Groves 2003; Scott and Csuti 1997). As we discussed, resilience implies something about the capacity to absorb or rebound from disturbance or change. We might focus on the most resilient or resistant habitats and populations first because they will have the best chance against future change and because they may also require a limited investment to secure (i.e., greatest benefit for the least cost). Representation refers to the distribution of biological and ecological diversity among populations or conservation units if there is opportunity to focus on more than one. We've considered the role of diversity in hedging, plasticity, and adaptation with disturbance and change. In this case, we might favor habitats or populations that would reflect distinctly different life history types or populations that exist in distinctly different environments under the assumption that they will respond in different ways to a changing environment that we cannot readily anticipate (e.g., Gibson and others 2009; Hilborn and others 2003; Lesica and Allendorf 1995). Redundancy represents the replication of conservation

efforts, units, or populations to minimize the probability that all will be lost in a single catastrophic event. There are obvious issues and tradeoffs between the number, size, and spatial distribution of replicates that are needed (e.g., many small versus a few large). Tradeoffs will depend largely on local conditions such as the frequency and extent of the dominant disturbance regime (e.g., the typical size of the stream networks influenced by fire and/or the length of channel segments influenced by debris flow); the dispersal distances possible for a given species; and the geographic gradients of diversity in genes, life history expression, and habitat potential. The general concepts, however, are that larger more productive populations have a better chance of surviving a given disturbance; multiple populations spread the risk that not all will be lost simultaneously; and diversity among those populations helps hedge against uncertainty and enhance the potential for adaptation to new conditions. A simple rule of thumb may be to seek a balance of all three concepts among the conservation units that are possible (Levin and Lubchenco 2008).

Favor actions robust to uncertainty.

Even as we learn more about the nature of climate change, substantial uncertainty remains. The world is warming, and patterns of precipitation and stream flow are changing in ways that we can anticipate but not always predict with precision. The changes have been and will become more dramatic and influential in some places than others. We will be surprised. The rate of change and the ultimate magnitude and variability of change will depend on the effectiveness of mitigation and myriad smaller scale controls and interactions that we can only guess at, while the conditions that influence those controls and interactions are variable and changing (Pielke 2009). This is not a new problem in the management of complex ecological systems. Management that depends on prediction of a sustainable harvest level or a threshold of acceptable habitat disruption simply has not proven very effective (Ludwig and others 1993; Poole and others 2004). The alternative is to accept that uncertainty and favor options that least depend on the precision of our knowledge or predictions (Lawler and others 2009; Ludwig and others 1993; Pielke 2009). Actions that generally contribute to the resilience of populations (e.g., removing barriers, expanding high-quality habitats, reducing non-climate stresses), regardless of the threat, might be favored over those that attempt to mitigate or control the influence of a specific threat (fire suppression or intentional barriers to protect small remnant populations; Fausch and others 2006). A corollary to this idea is to favor the conservation or restoration of natural process over the restoration of structure (Beechie and others 2008, 2010) or the control of processes that require a continuous investment of time, energy, and money to maintain (Rieman and others 2010). Joyce and others (2009) term this strategy one of "no regrets," implying that the choices are likely to provide important benefits regardless of the ultimate effects of climate change.

The bottom line is that there are no strict rules for prioritization but it remains important, if not critical. Random actions are more likely to waste time and limited resources and to be less effective than those focused through a specific and strategic process that is relevant to the system at hand. The considerations outlined above may provide some foundation, but ultimately, managers and biologists will need to devise a strategy that makes sense for the systems they know. Further discussion and some examples are available in other work (e.g., Beechie and others 2006; Fausch and others 2006; Frissell and others 1997; Groves 2003; Roni and others 2002). For the most part, the work we can do is not new. The tools of watershed, habitat, species, and population conservation and restoration are well known to most biologists working throughout the Rocky Mountain West. In the context of climate change, however, we have an even greater need to implement those tools as effectively and efficiently as possible.

Facilitate Transitions to New Conditions

It seems likely, given current and anticipated trends in climate change, that it simply will not be possible to conserve all existing species' populations or the structure of aquatic communities across the streams and habitats where they currently occur. In some cases, the existing habitats suitable for some species and communities we hope to conserve will be lost locally or they may be shifted upstream or north. Many terrestrial organisms may be able to move to other suitable habitats or track the changing distribution of habitats. But for others, and especially for fishes confined to stream networks constrained by natural and anthropogenic barriers, that may be impossible. As a result, there has been considerable discussion and debate in the conservation literature on "facilitated dispersal" or "assisted migration" (e.g., Ricciardi and Simberloff 2009; Sax and others 2009; Vitt and others 2009). Concerns exist because introductions of non-native species can create a suite of ecological and environmental problems (Ricciardi and Simberloff 2009). Others argue that facilitation should only mimic transitions or dispersals that would have occurred naturally but are now

impossible or too slow because of anthropogenic fragmentation of existing habitats (Vitt and others 2009). In reality, fisheries biologists and managers have long facilitated dispersal and migration of fishes through hatchery supplementation programs, installation of passage structures at natural or human-caused barriers, trap and haul programs (e.g., Columbia River salmon), and intentional species introductions. There are even significant efforts to move fishes like Gila trout to artificial refugia (i.e., hatcheries) and back to historical or new habitats in response to wildfire-driven catastrophic disturbances (Brooks 2006). In each case, managers must weigh the costs and potential risks both of attempting to control ecological processes that no longer occur naturally and of establishing non-native communities that may function without further substantial investment (Fausch and others 2006). We do not argue for any particular solution but suggest that those decisions be placed in the context of climate change. In streams and habitats where it seems unlikely that native species and communities will persist in the future, managers may choose to facilitate the transition to a new community actively (e.g., through introduction of new species to the existing habitat or of the old species to a new habitat) or passively (e.g., through removal of barriers that exclude invasions). Consideration of the sensitivity of habitats, populations, and communities outlined in the previous section could provide important context here as well, although the implied options are different. As suggested above, it will be important to consider the ecological, evolutionary, and social values involved since they directly define the tradeoffs.

Develop Local Information

Developing and implementing efficient adaptation and facilitation strategies for climate change requires information specific to local climatic trends and integration of this knowledge with current and future status of local landscapes, streams, and aquatic resources of concern. Global climate models (GCM), though capable of providing climatic predictions for all areas of the Earth, have minimum resolutions of hundreds to thousands of square kilometers and were not designed to provide the sort of site-specific information typically required for ecological assessments (Wiens and Bachelet 2009). Moreover, GCM outputs usually consist only of air temperature and precipitation, which must be translated to relevant habitat features for aquatic biotas. Fortunately, new analytical techniques for streams; increased availability of large, geo-referenced databases of aquatic attributes; online summaries of local climate information; and advances in spatial technologies like geographic information systems (GIS) and remote sensing can provide managers a potential wealth of site-specific information and integration tools. Linking this information with local fish and habitat surveys provides a powerful means of understanding climate-related trends and risks to individual populations or other aquatic resources. In this section, we highlight factors, techniques, and data sources to consider when developing local climatic assessments.

Understand context. At the outset of any local climatic assessment, it is important to understand the broader context associated with regional climate trends, stream responses, and species' distributions. This knowledge provides context that may help managers narrow their focus to resources that are most at risk (Hurd and others 1999; Wiens and Bachelet 2009). Numerous studies that document both observed and projected trends across the western United States are useful for understanding this context. For example, research that links mid-winter air temperatures to changing hydrology and flood risks (figure 10; Hamlet and Lettenmaier 2007) can highlight specific watersheds or forests where increases in winter flooding might be anticipated and others where flooding may decrease. Similarly, earlier and smaller snowmelt runoff, when accompanied by reductions in annual precipitation, means that summer low flows are decreasing rapidly in some areas (Luce and Holden 2009) but slowly in others (Regonda and others 2005; Stewart and others 2005). Managers may be relatively certain about changes where historical trends agree with future climate model projections, but in other areas, especially in the Southwest, projections sometimes differ from recent historical observations and additional uncertainty exists (Hoerling and Eischeid 2007; Mote and others 2005; GCRP 2009). Finally, a growing number of regional bioclimatic and conservation assessments now exist that provide population inventories and projections of future climate effects across significant portions of species' ranges (e.g., Battin and others 2007; Keleher and Rahel 1996; Rieman and others 2007; Young 2008). These are useful for identifying the relative position of local populations within the range of future climate changes.

Synthesize existing information. Many types of data exist that can be compiled into a local climatic assessment. Examples of useful information are: fish survey data, stream temperature records, and information on watershed disruption or habitat loss. The latter includes any activity that potentially impairs riparian, watershed, or stream integrity. Of particular interest in many areas will be inventories of road culverts and water diversion structures that create fish barriers or warm stream reaches with degraded banks or riparian areas, as well as channels that are prone to debris flows or subject to chronic road sedimentation. These features or locations often provide some of the best opportunities for reducing existing stresses (Fausch and others 2006; Hendrickson and others 2008).

Useful fish data include spawn timing, dates of fry emergence, and migration times of adults or juveniles past weirs or other structures where they are counted (e.g., Crozier and others 2008; Juanes and others 2004). If these surveys are repeated over time, they may reveal trends that show how fish populations are responding to environmental change (Elliot and others 2000). Even a few years of data from contrasting climate years, however, could provide valuable clues regarding longer-term responses given the plasticity of phenotypic traits. Spatially distributed fish population surveys and stream temperature data are often abundant and useful for building bioclimatic models that facilitate inference beyond sample dates and locations and can be used to understand landscape, stream, and climatic factors that regulate species distributions and aquatic thermal regimes. When compiling these data, potential sources in other Federal or State agencies should be considered and may provide useful supplements to data collected on National Forests. Any data sources require careful checks for methodological consistency, but regional databases that are compiled using standardized protocols and geo-referenced survey sites are becoming more common (e.g., Meyer and others 2009; Pont and others 2009). Stream temperature data are now routinely collected by many resource agencies and municipalities, and digital recording devices provide consistent data quality (Dunham and others 2005). Increasingly, these data are being archived and made available through corporate databases like the Aquatic Surveys module in the National Resource Information System (http://www.fs.fed.us/emc/nris/), the Pacfish-Infish Biological Opinion monitoring group (http://www.fs.fed.us/biology/fishecology/emp/index.html), and the National Water Information System (http://waterdata.usgs.gov/nwis).

Other useful information such as hydrography datasets and digital terrain models can be obtained at differing resolutions and can be used in a GIS environment to represent a study area and describe important environmental gradients. The National Hydrography Dataset–Plus (http://www.horizon-systems.com/nhdplus/), available at a scale of 1:100,000, provides a suite of applications and attributes calculated for individual stream segments from the National Elevation Dataset, the National Land Cover Dataset, and the Watershed Boundary Dataset. Climatic attributes such as interpolated air temperatures and flow values based on nearby stream gaging sites are included as stream attributes but represent only static values averaged over time. Understanding and calculating local climate trends requires obtaining historical climate data for air temperature and precipitation from weather stations maintained by the National Oceanic and Atmospheric Administration (http://www.ncdc.noaa.gov/oa/ncdc.html) or the Snowpack Telemetry sensor network (http://www.wcc.nrcs.usda.gov/snow/) and stream flow data from U.S. Geological Survey gage sites (http://waterdata.usgs.gov/nwis/; Falcone and others 2010). Summary metrics that describe annual timing, magnitude, and variance in these factors are easily calculated from the raw data for stations maintained in or near the area of interest and can be plotted relative to time to provide site-specific descriptions of historic climatic trends. Descriptions of how these trends translate to elevations necessary for maintaining a given temperature isopleth may be calculated using a web tool called the North American Freezing Level Tracker (http://www.wrcc.dri.edu/cwd/products/). Extrapolating historical trends into the future may indicate when and by how much local climatic conditions could change, although this simple approach can be biased given the length of the time series and changes in the rates of climate responses (IPCC 2007). Regional climate centers like the Western Regional Climate Center (http://www.wrcc.dri.edu/) and the Climate Impacts Group in Seattle, Washington, (http://cses.washington.edu/cig/) also provide many valuable climate summaries and tools for analyzing historical and future trends at local and regional scales.

Model to fill gaps. Organizing existing data inventories often reveals a wealth of useful information but also highlights many data gaps and limitations associated with raw data summaries. Where gaps exist, new data may be collected, but in many instances, additional data collections may be too costly or take too long, and models built from existing databases can be used to make predictions in unsampled locations. These models can be used to interpolate among sites and to predict future habitat and fish distributions under warmer climates. An entire subdiscipline focused on "bioclimatic" models now exists within the scope of climate-related ecological work (Dormann and others 2007; Guisan and Thuiller 2005; Latimer and others 2006). Bioclimatic

models are similar to traditional species distribution or habitat models but integrate climate variables to the suite of predictive covariates. This establishes a linkage between current species distributions or habitat conditions and climatic factors that can be used to predict future conditions by changing the climate variable input values. This approach works well as long as the historical correlation between climatic predictors and a response variable remains consistent, but these relationships may also change through time due to unforeseen events (e.g., invasion by an exotic species) or an incomplete understanding of the system being modeled (i.e., omission of important predictor variables). Another major source of uncertainty is selecting the correct parameter values to represent future conditions, as significant unknowns exist regarding the Earth's climate trajectory. Still, bioclimatic models provide a means for assessing a range of possible futures. This ability is important for assessing relative risk across populations or other valued resources and can help distinguish among potential lost causes, core areas that are more resistant and resilient, and areas intermediate to the extremes where management interventions might play a decisive role in the outcome.

Examples of bioclimatic models developed for salmonids in the Rocky Mountains are Keleher and Rahel's (1996) work on projecting changes in thermally suitable habitat for a suite of trout species, and a more recent study by Rieman and others (2007) that examined potential changes in bull trout distributions across the interior Columbia River basin (figure 18). Characteristic of bioclimatic studies, both were of broad spatial extent and limited local precision—a compromise that has often been necessary given previous analytical capabilities and a forced reliance on crude approximations of local climatic conditions (e.g., elevation and air temperature as surrogates for stream temperature). Previous bioclimatic models, therefore, were generally best used to provide strategic overviews of potential regional changes, with guidance at smaller spatial scales generally lacking.

These limitations are beginning to disappear with the compilation of large stream databases, accurate spatial referencing from global positioning systems, and improvements in GIS and remote sensing that facilitate better measurements of environmental characteristics. Synthetic stream hydrographs can now be generated for historic and future scenarios throughout river networks using a variety of hydrologic models when linked to climate model projections and digital representations of landscapes (Battin and others 2007; Hamlet and Lettenmaier 2007). The Variable Infiltration Capacity

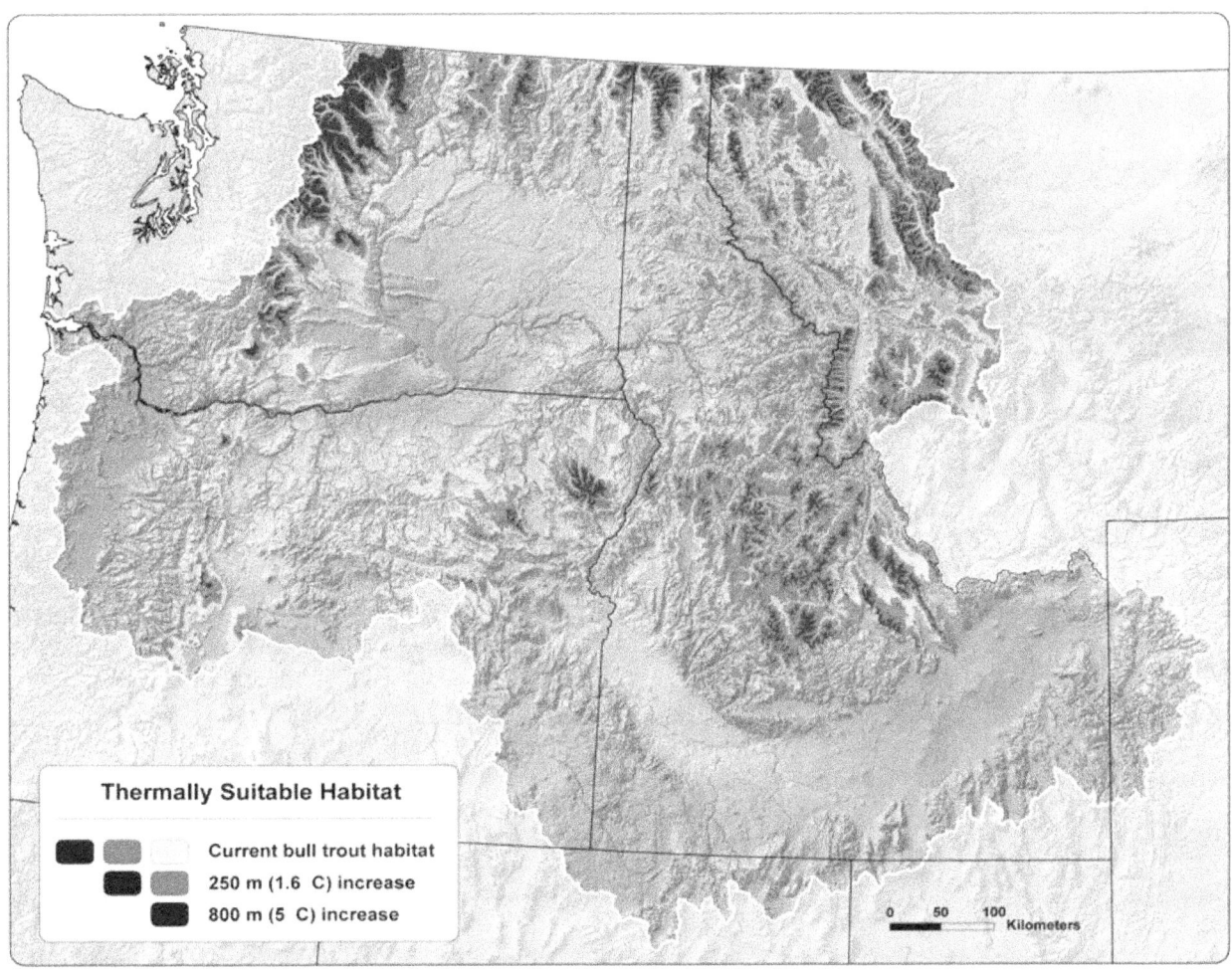

Figure 18. Changes in thermally suitable habitat under different air temperature increases for bull trout spawning and juvenile rearing within the Interior Columbia River basin predicted by a bioclimatic model. All shades of blue indicate historically suitable habitats; whereas lighter shades of blue indicate habitats that could be lost with future air temperature warming of 1.6 °C to 5 °C (from Rieman and others 2007).

(VIC) model (Hamlet and Lettenmaier 2007; Liang and Lettenmaier 1994), in particular, has been widely applied across much of the western United States to model climate change scenarios (e.g., Christensen and others 2004; Hamlet and Lettenmaier 2007). Recently, Wenger and others (2010) adapted the VIC model to work in small headwater streams and derived a series of ecologically relevant flow metrics for use in bioclimatic models. Improved statistical models that account for the spatial structure of streams, either by treating them as networks (Peterson and others 2007; Ver Hoef and Peterson 2010) or hierarchical systems (Cressie and others 2009), have also been developed and substantially improve predictive accuracy by accommodating spatial autocorrelation among sample locations (e.g., Isaak and others 2010; Rieman and others 2006). Using these new statistical techniques in conjunction with a large stream temperature database (n = 780) compiled from several

resource agencies, Isaak and others (2010) built stream temperature models that explained 93 percent of the variation in temperatures over a 13-year period across an extensive river network. Similar results have been achieved using spatial stream techniques with other water quality attributes (Gardner and McGlynn 2009; Peterson and Urquhart 2006) and fish population attributes (Peterson and Ver Hoef 2010), and those results suggest a future wherein bioclimatic relationships in streams can be examined in great detail (McIntire and Fajardo 2009).

As our ability to model aquatic ecosystems improves, distributions of fish species and important habitat features will be predicted at resolutions that enable managers to make useful comparisons among different management options (figure 19). Some choices regarding resource allocations may be obvious, especially if a habitat or population appears likely to disappear (or

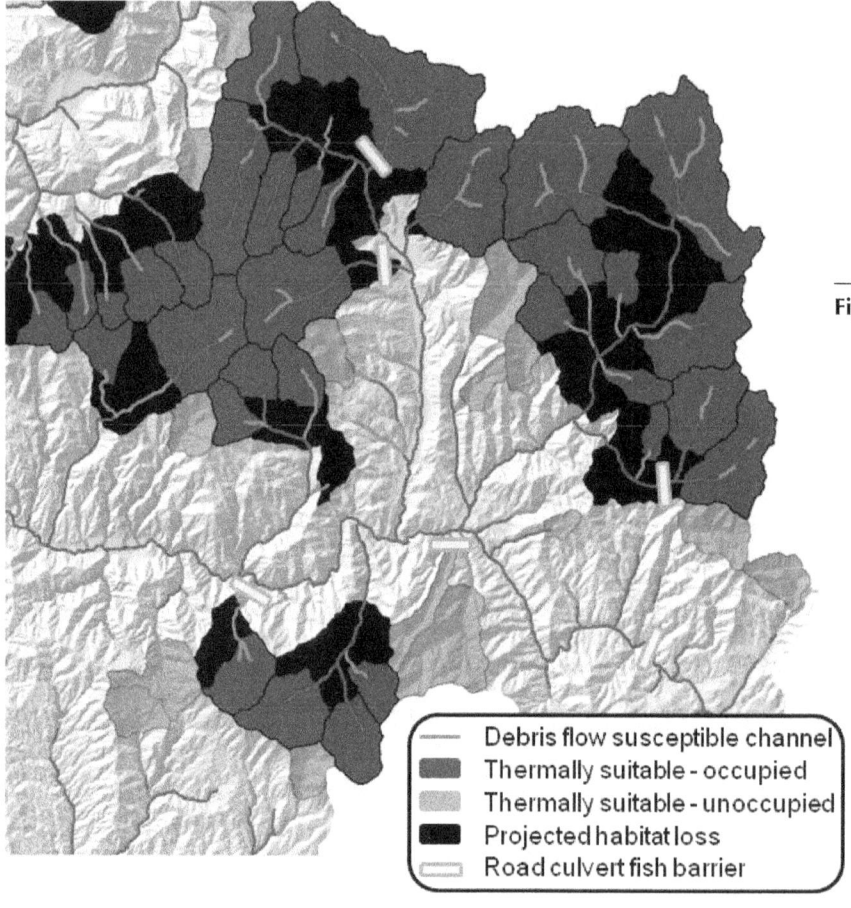

Figure 19. Hypothesized scenario depicting landscape-level effects of climate change on headwater habitats for a sensitive fish species. Black areas depict currently suitable habitats that will become thermally unsuitable by mid-century; red lines indicate channels susceptible to debris flows if a wildfire occurs; and yellow bars represent road culverts that block fish passage. Which culvert barriers should be given highest priority for improving fish passage if sufficient resources are not available to modify all barriers?

--- Debris flow susceptible channel
■ Thermally suitable - occupied
▨ Thermally suitable - unoccupied
■ Projected habitat loss
▭ Road culvert fish barrier

persist) under most future climate scenarios. In many cases, however, choices will be less obvious due to complex interactions among multiple factors. Streams or populations that fall into this grey area are where management interventions may be instrumental in tipping the balance toward desired outcomes.

Monitor to document trends. Regardless of our ultimate ability to predict the biophysical attributes of streams, differences will invariably arise between the predicted and observed future states of these systems. Monitoring programs that are targeted at key biophysical attributes will be essential for validating and refining model projections and determining real rates of change. Due to costs, extensive monitoring programs may not be feasible for all attributes of potential interest; but stream temperatures, stream discharge, and fish distribution are likely to be common priorities.

Stream temperature measurements are easily and reliably obtained and are relatively inexpensive using modern digital sensors (Dunham and others 2005; Isaak and others 2010). Temperature sensor networks can be deployed in a few days or weeks to provide inference ranging in scale from individual streams to river networks or entire basins. Basic recommendations for the design of temperature sensor networks in streams are available (Isaak and others 2009; http://www.fs.fed.us/rm/boise/AWAE/projects/stream_temperature.shtml), although site-specific applications will vary with local objectives. A few years of temperature data from a sensor network is sufficient to start identifying streams that may be more (or less) sensitive to heating based on inter-annual differences in climatic conditions. Several years of annual stream temperature from a site may be enough to reconstruct the long-term thermal history at that site (or future projections) based on relationships with air temperatures measured at nearby weather stations (Mohseni and others 2003; Mohseni and Stephan 1999). Such reconstructions may help compensate for the short length of stream temperature monitoring in many areas.

Stream discharge has long been monitored by the U.S. Geological Survey, but most gage sites are located on larger streams and rivers that provide major water supplies. The relative dearth of flow measurements on smaller streams has, in part, been limited by cost, but new instruments based on pressure transducers are reducing logistical and financial costs and could open a new era of flow monitoring. Similar to stream temperatures, historical flow reconstructions may be possible with a few years of data by relying on correlations with long-term records at nearby gages. Even longer perspectives (i.e., several hundred to a thousand years) are possible from dendrochronology that uses tree ring width as a proxy for precipitation and stream flow (e.g., Graumlich and others 2003). Although these reconstructions sacrifice temporal resolution for extended length, they provide an important historical context for contemporary patterns.

Data on fish populations may be relatively abundant, but most often consist of spatially distributed information from reach-based surveys rather than long-term monitoring at individual sites. An exception may occur where weirs, screw-traps, or other means of counting fish during seasonal migrations have been maintained through time. Indeed, most of the direct evidence of climate change on fish populations is derived from trends in the timing of anadromous spawning migrations during the latter half of the twentieth century (e.g., Crozier and others 2008; Quinn and Adams 1996; Robards and Quinn 2002). Case histories that document climate-induced shifts in spatial distributions are almost non-existent, despite these shifts being the most common prediction generated by bioclimatic assessments (e.g., Keleher and Rahel 1996; Rieman and others 2007). Hari and others (2006) provide one of the few examples that shows brown trout populations in Switzerland have shifted upstream in recent decades associated with steadily increasing temperatures and frequency of disease. Monitoring to detect similar shifts in populations and communities of western fishes will be crucial to understand actual rates of habitat loss. Weak populations in thermally marginal areas near a species' downstream or upstream distributional extent are expected to be the most sensitive to climate effects and could be specifically targeted to provide this information (e.g., Isaak and others 2009; Rieman and others 2006).

Depending on the area of interest and available resources, it may or may not be possible for a single entity to develop and maintain adequate monitoring programs. The Pacfish-Infish Biological Opinion Monitoring Program and Environmental Monitoring and Assessment Program (Herlihy and others 2000; Roper and others 2010) are examples of single agency initiated efforts that have proven their utility through time. Between National Forests and State and Tribal agencies, the potential exists for significant redundancies.

Such inefficiencies are problematic when faced with the scale and resolution of monitoring that climate change may necessitate. Interagency (or inter-forest) efforts to coordinate monitoring, therefore, will likely be important (see below). Standardization of data collection techniques; coordination of sample locations; and use of centralized, integrated databases could provide significant leverage for local monitoring efforts, decrease costs, and improve the ability to track environmental change.

Coordinate Efforts

Climate change represents a major challenge to aquatic biologists and managers. Managers of other natural resources face similar issues. Because aquatic systems are directly influenced by the conditions and management of other resources and by values and actions of society at large, aquatic management responses to climate change are not likely to be very effective if they are not coordinated with other efforts to understand and adapt. There are at least three areas where that coordination seems particularly important: across natural resource disciplines, particularly those focused on terrestrial and aquatic systems; across agencies and organizations; and between natural resource specialists and the general public.

Across disciplines. The history of land use management has often been one of conflict among disciplines or distinct natural resource objectives (e.g., timber versus fish). Biologists and managers who are focused on aquatic or terrestrial issues linked to wildfire, for example, may view the other's efforts as a constraint on their own progress (e.g., Rhodes and Baker 2008; Rieman and others 2003; Rieman and others 2010). Conflict is accentuated by multiple jurisdictions, differences in mandates and subsequent goals or objectives, mismatch in temporal and spatial scales of issues, limited or ineffective communication, and limited management resources (Naiman and others 1998; Noss and others 2006; Rieman and others 2003). The results have often been attempts to reconcile differences through piecemeal negotiation or regulatory processes that occur project by project (Baron and others 2002), with patchwork integration of individual resource management plans after the fact or with political solutions that favor one option or the other depending on the current administrative rules and direction (Rieman and others 2003). The inextricable linkages between terrestrial and aquatic systems and the simultaneous disruption of both through past land use activities means that opportunities for more effective coordination of terrestrial and aquatic restoration and conservation could be widespread (Rieman and others 2000; Rieman and others 2010). More successful collaboration might be built on broader perspectives that examine potential opportunities and conflict across entire landscapes (Noss and others 2006; Rieman and others 2010); active communication based on clear definition of values, goals, and objectives early in the process (Naiman and others 1998; Firth 1998); and a focus on the long-term restoration of ecological processes that contribute to the resilience of both terrestrial and aquatic systems (Rieman and others 2010).

Across agencies or jurisdictions. The conservation and management of natural resources dependent on, or directly influencing, aquatic systems fall under many different jurisdictions and across agencies and authorities with different missions, histories, and capacities. A broad collection of non-governmental organizations often work to facilitate or restrict different elements of that mix, as well. A common thread is that much, if not all, of the natural resource values and services important to each will be influenced by climate change in some way. Basic climate science has expanded rapidly in recent years, and substantial efforts are now focused on refining and downscaling GCMs to better understand the effects at finer resolutions approaching the scale of local management decisions. Researchers are actively developing linked hydrologic, biological, and ecological models to explore species and ecosystem responses (e.g., Isaak and others 2010; Wenger and others 2010). New tools and multiple syntheses are being developed to help managers sort through the complexity and potential implications and weigh alternatives.

In short, the science and information potentially useful to managers confronted with climate change is rapidly expanding. It is difficult for any one person to keep up with the literature and development in a single field such as aquatic ecology and even more challenging (if not impossible) in the broader interdisciplinary suite that is important to integrated landscape management. Fortunately, there are initiatives to summarize, synthesize, and serve information as it develops (e.g., The Forest Service Climate Change Resource Center; http://www.fs.fed.us/ccrc/). Interdisciplinary and interagency consortia are being discussed to extend capacity for modeling and analysis, leverage funding, minimize duplication, and potentially extend the communication across resource disciplines considered above. Biologists and managers struggling with climate change might find these efforts helpful, but they might also create other opportunities to extend limited resources and capacity. The Forest Service, State fish and game agencies, and State water quality biologists and managers, for example, might all benefit from high-resolution downscaled hydrologic predictions or regional stream temperature monitoring networks. Sharing the cost to develop and serve such key information with appropriate local resolution could be useful. Designation of a "climate lead" in each agency, work group, management office, or region might help facilitate discussion, track and communicate developments, and facilitate new linkages within and across agencies.

With the public. Ultimately, management of natural resources and any response to climate change must be responsive to the values society places on those resources and the tradeoffs required in the process. The effects of climate change on aquatic systems will probably be aggravated by continued development and land conversion and by growing demands for increasingly limited water. More frequent or more extreme disturbances in the form of wildfire, flooding, and drought could lead to more extensive attempts to mitigate effects through more extensive forest thinning and harvest, diking, and water storage, respectively. Where human lives, property, and livelihoods are concerned, natural resource issues will be considered only secondarily, if at all, unless they can be clearly linked to more pragmatic benefits (such as water storage and flood attenuation) that healthy, functioning watersheds provide in addition to maintenance of biological diversity, fishes, and fisheries. Paul Angermeier (2007) has argued (convincingly, in our view) that public education on aquatic issues and the tradeoffs we make with other values may be the single most important thing aquatic managers and biologists can do for the long-term conservation of aquatic ecosystems, biological diversity, fishes, and fisheries. Biologists and managers working in and with local communities may have the best opportunity and understanding to do that effectively. The realities of climate change and the broader political and economic challenges could make that even more important.

The bottom line is that climate change will influence virtually all natural resource management efforts and values in important ways. Few, if any, managers and biologists will have the capacity to understand or respond without reaching out to, guiding, and learning from others who influence their activities and decisions, the availability of critical information, and their opportunities to move forward.

Epilogue

Understanding and responding to climate change requires an ongoing dialogue between management and research that is informed by observed environmental trends. There has been a flood of new information to consider based on climate models and early downscaling with projections to a host of natural resource and social issues, including conservation of native fishes. Those models will be refined and expanded and they will provide important insight. Models must be tested against reality, however, so one of the most important foundations for a continuing dialogue will be syntheses and summaries of trends in monitoring data. Most of what is known about climate change came from long-term monitoring that was largely in place before climate became a central issue. That understanding will be extended by efforts to form cross-disciplinary and interagency collaborations and more complex infrastructure, including the databases and continued and refined monitoring, modeling, and decision support tools needed to consider the potential alternatives for adaptation across regions and river basins.

Those broad initiatives will depend on the integration of regional and local information. Like many others, our initial interest in climate change came from a growing discussion in the general literature. That interest became focused as we considered systems and species we worked with in the context of climate change and the results from our own monitoring and long-term research. As we developed our data, the trends and projections that emerged made climate change real to us. That led to new questions and more focused work. From that experience, we believe local monitoring efforts with climate change in mind, even if limited to a single temperature sensor maintained in a single stream, will be among the most important steps biologists and managers can take in response to climate change. As information develops, biologists and managers will help test hypotheses relevant to the systems they know. That will point out important strengths and weaknesses and will help refine projections of how aquatic systems will or won't change in the future. Biologists and managers will gain better insight to the utility of some management actions and the futility of others. Empirical observations may also provide a basis for making difficult decisions. For example, redirecting conservation efforts away from marginal populations of a sensitive species based solely on climate model projections would, at present, be difficult to justify. If, however, model projections were accompanied by data from rigorous monitoring that had documented previous climate-related population losses, such decisions might become more acceptable.

Early model projections, despite their utility, are fraught with uncertainties, should be viewed critically, and must be revised periodically as new information becomes available. The single largest source of uncertainty is simply how much and how fast the Earth's climate will warm. Other important uncertainties, however, include the specifics of how future warming will affect freshwater ecosystems, western streams, and aquatic communities. Researchers and managers can generate much of the needed information independently, but partnerships between research and management could be particularly useful.

Research is needed to develop models that translate both current and future climate effects to aquatic ecosystems, but data to build and validate such models will be required across broad spatial and temporal scales. Research likely will never have the resources to do that work independently but could harness data from routine monitoring programs. Monitoring of key biophysical attributes, therefore, may provide not only trend data but also much of the information necessary to develop models, test hypothesized mechanisms, or stimulate new research by documenting unanticipated responses. A collaborative, adaptive system that is characterized by iterations of model development/predictions and subsequent calibration/refinement against monitoring data and a record of management results could systematically reduce uncertainties.

As new information regarding the Earth's climate trajectory and its effects on aquatic systems develops, model projections and management implications will evolve. Model projections will become more accurate as understanding of key processes improves and supporting databases expand. Spatially explicit forecasts of habitat conditions, species distributions, and probabilities of persistence for the latest climate scenarios will be possible, much the way that harvest quotas are forecast for commercial fisheries. Instead of revising forecasts over intervals of weeks and months that are relevant to a fishery, however, forecasts of climate change effects might be updated over longer intervals (e.g., 5 to 10 years). Management decisions will have to be informed of, and open to, the evolving state of knowledge. Partnerships between research and management and continued monitoring through both local and broader collaborative efforts could help that process and promote effective adaptation to a changing climate.

References

Abatzoglou, J. T.; Redmond, K. T. 2007. Asymmetry between trends in spring and autumn temperature and circulation regimes over western North America. Geophysical Research Letters. 34(L18808): doi: 10.1029/2007GL030891.

Allendorf, F. W.; Bayles, D. W.; Bottom, D. L.; Currens, K. P.; Frissell, C. A.; Hankin, D.; Lichatowich, J. A.; Nehlsen, W.; Trotter, P. C.; Williams, T. H. 1997. Prioritizing Pacific salmon stocks for conservation. Conservation Biology. 11: 140-152.

Angermeier, P. L. 2007. The role of fish biologists in helping society build ecological sustainability. Fisheries. 32(1): 9-20.

Angermeier, P. L.; Neves, R. J.; Kaufman, J. W. 1993. Protocol to rank value of biotic resources in Virginia streams. Rivers. 4: 20-29.

Barnett, T. P.; Pierce, D. W.; Hidalgo, H. G.; Bonfils, C.; Santer, B. D.; Das, T.; Bala, G.; Wood, A. W.; Nozawa, T.; Mirin, A. A.; Cayan, D. R.; Dettinger, M. D. 2008. Human induced changes in the hydrology of the western United States. Science. 319: 1080-1083.

Baron, J. S.; Poff, N. L.; Angermeier, P. L.; Dahm, C. N.; Gleick, P. H.; Hairston, N. G. J.; Jackson, R. B.; Johnston, C. A.; Richter, B. D.; Steinman, A. D. 2002. Meeting ecological and societal needs for freshwater. Ecological Applications. 12(5): 1247-1260.

Bartholow, J. M. 2005. Recent water temperature trends in the lower Klamath River, California. North American Journal of Fisheries Management. 25: 152-162.

Battin, J.; Wiley, M. W.; Ruckelshaus, M. H.; Palmer, R. N.; Korb, E.; Bartz, K. K.; Imaki, H. 2007. Projected impacts of climate change on salmon habitat restoration. Proceedings of the National Academy of Sciences (USA). 104: 6720-6725.

Beechie, T.; Buhle, E.; Ruckelshaus, M.; Fullerton, A.; Holsinger, L. 2006. Hydrologic regime and the conservation of salmon life history diversity. Biological Conservation. 130: 560-572.

Beechie, T.; Pess, G.; Roni, P.; Giannico, G. 2008. Setting river restoration priorities: a review of approaches and a general protocol for identifying and prioritizing actions. North American Journal of Fisheries Management. 28: 891-905.

Beechie, T.; Sear, D. A.; Olden, J. D.; Pess, G. R.; Buffington, J. M.; Moir, H.; Roni, P.; Pollock, M. 2010. Process-based principles for restoring river ecosystems. BioScience. 60: 209-222.

Beever, E. A.; Ray, C.; Mote, P. W.; Wilkening, J. L. 2010. Testing alternative models of climate-mediated extirpations. Ecological Applications. 20: 164-178.

Bilby, R. E.; Bisson, P. A.; Coutant, C. C.; Goodman, D.; Gramling, R.; Hanna, S.; Loundenslager, E.; McDonald, L.; Philipp, D.; Riddell, B. 2003. A review of strategies for recovering tributary habitat. ISAB 2003-2. Portland, OR: Independent Scientific Advisory Board for the Northwest Power Planning Council, the National Marine Fisheries Service, and the Columbia River Basin Indian Tribes. 54 p.

Bisson, P. 2008. Salmon and trout in the Pacific Northwest and climate change. [Online]. U.S. Department of Agriculture, Forest Service, Climate Change Resource Center. Available: http://www.fs.fed.us/ccrc/topics/salmon-trout.shtml [June 16, 2008].

Bisson, P. A.; Dunham, J. B.; Reeves, G. H. 2009. Freshwater ecosystems and resilience of Pacific salmon: habitat management based on natural variability. [Online]. Ecology and Society. 14: 45. Available: http://www.ecologyandsociety.org/vol14/iss41/art45/

Bisson, P. A.; Rieman, B. E.; Luce, C.; Hessburg, P. F.; Lee, D. C.; Kershner, J. L.; Reeves, G. H.; Gresswell, R. E. 2003. Fire and aquatic ecosystems of the western USA: current knowledge and key questions. Forest Ecology and Management. 178(1-2): 213-229.

Blum, M. D.; Tornqvist, T. E. 2000. Fluvial responses to climate and sea-level change: a review and look forward. Sedimentology. 47: 2-48.

Bottrill, M. C.; Joseph, L. N.; Carwardine, J.; Bode, M.; Cook, C.; Game, E. T.; Grantham, H. S.; Kark, S.; Linke, S.; McDonald-Madden, E.; Pressey, R. L.; Walker, S.; Wilson, K. A.; Possingham, H. P. 2009. Finite conservation funds mean triage is unavoidable. Trends in Ecology and Evolution. 24: 183.

Boxall, G. D.; Giannico, G. R.; Li, H. R. 2008. Landscape topography and the distribution of Lahontan cutthroat trout (*Oncorhynchus clarki henshawi*) in a high desert stream. Environmental Biology of Fishes. 82: 71-84.

Bozek, M. A., Young, M. K. 1994. Fire mortality resulting from delayed effects of fire in the greater Yellowstone ecosystem. Great Basin Naturalist. 54:91-95.

Brannon, E. L.; Powell, M. S.; Quinn, T. P.; Talbot, A. 2004. Population structure of Columbia River basin Chinook salmon and steelhead trout. Reviews in Fisheries Science. 12: 99-232.

Brenkman, S. J.; Corbett, S. C.; Volk, E. C. 2007. Use of otolith chemistry and radiotelemetry to determine age-specific migratory patterns of anadromous bull trout in the Hoh River, Washington. Transactions of the American Fisheries Society. 136: 1-11.

Breshears, D. D.; Cobb, N. S.; Rich, P. M.; Price, K. P.; Allen, C. D.; Balice, R. G.; Romme, W. H.; Kastens, J. H.; Floyd, M. L.; Belnap, J.; Anderson, J. J.; Myers, O. B.; Meyer, C. W. 2005. Regional vegetation die-off in response to global-change-type drought. Proceedings of the National Academy of Sciences. 102: 15144-15148.

Breshears D. D.; Myers, O. B.; Meyer, C. W.; Barnes, F. J.; Zou, C. B.; Allen, C. D.; McDowell, N. G.; Pockman, W. T. 2009. Tree die-off in response to global change-type drought: mortality insights from a decade of plant water potential measurements. Frontiers in Ecology and the Environment. 4: 185-189.

Brooks, J., ed. 2006. Emergency evacuation procedures for Gila trout. Gila trout recovery team, U.S. Department of the Interior, Fish and Wildlife Service; Arizona Game and Fish Department, New Mexico Department of Game and Fish; U.S. Department of Agriculture, Forest Service; University of New Mexico.

Brown, D. K.; Echell, A. A.; Propst, D. L.; Brooks, J. E.; Fisher, W. L. 2001. Catastrophic wildfire and number of populations as factors influencing risk of extinciton for Gila trout (*Onchorhnchus gilae*). Western North American Naturalist. 61: 139-148.

Brown, T. C.; Hobbins, M. T.; Ramirez, J. A. 2008. Spatial distribution of water supply in the coterminous United States. Journal of the American Water Resources Association. 44: 1474-1487.

Chatters, J. C.; Butler, V. L.; Scott, M. J.; Anderson, D. M.; Neitzel, D. A. 1995. A paleoscience approach to estimating the effects of climatic warming on salmonid fisheries of the Columbia River basin. Pages 489-496 in: Beamish, R. J., ed. Climate Change and Northern Fish Populations. Canadian Journal Special Publication Fisheries and Aquatic Sciences. 121.

Christensen, N. S.; Wood, A. W.; Voisin, N.; Lettenmaier, D. P.; Palmer, R. N. 2004. The effects of climate change on the hydrology and water resources of the Colorado River basin. Climatic Change. 62: 337-363.

Climate Change Science Program (CCSP). 2009: Thresholds of cimate change in ecosystems. A report by the U.S. Climate Change Science Program and the Subcommittee on Global Change Research [Fagre, D.B., C.W. Charles, C.D. Allen, C. Birkeland, F.S. Chapin III, P.M. Groffman, G.R. Guntenspergen, A.K. Knapp, A.D. McGuire, P.J. Mulholland, D.P.C. Peters, D.D. Roby and George Sugihara]. Synthesis and Assessment Product 4.2. Reston, VA: U.S. Geological Survey. 157 p.

Coleman, M. A.; Fausch, K. D. 2007. Cold summer temperature limits recruitment of age-0 cutthroat trout in high elevation Colorado streams. Transactions of the American Fisheries Society. 136: 1231-1244.

Connor, W. P.; Sneva, J. G.; Tiffan, K. F.; Steinhorst, R. K.; Ross, D. 2005. Two alternative juvenile life history types for fall Chinook salmon in the Snake River basin. Transactions of the American Fisheries Society. 134: 291-304.

Coutant, C. C. 1999. Perspectives on temperature in the Pacific Northwest's freshwaters. ORNL/TM-1999/44. Oak Ridge, TN: Oak Ridge National Laboratory.

Cressie, N.; Calder, C. A.; Clark, J. S.; ver Hoef, J. M.; Wikle, C. K. 2009. Accounting for uncertainty in ecological analysis: the strengths and limitations of hierarchical statistical modeling. Ecological Applications. 19: 553-570.

Crozier, L. G.; Hendry, A. P.; Lawson, P. W.; Quinn, T. P.; Mantua, N. J.; Battin, J.; Shaw, R. G.; Huey, R. B. 2008. Potential responses to climate change in organisms with complex life histories: evolution and plasticity in Pacific salmon. Evolutionary Applications. 1: 252-270.

Crozier, L. G.; Zabel, R. W.; Hockersmith, E. E.; Achord, S. 2010. Interacting effects of density and temperature on body size in multiple populations of Chinook salmon. Journal of Animal Ecology. 79: 342-349.

Daly, C.; Conklin, D. R.; Unsworth, M. H. 2009. Local atmospheric decoupling in complex topography alters climate change impacts. International Journal of Climatology. doi:10.1002/joc.2007.

Danehy, R. J.; Colson, C. G.; Duke, S. D. 2010. Winter longitudinal thermal regime in four mountain streams. Northwest Science. 84: 151-158.

Dormann, C. F. 2007. Promising the future? Global change projections of species distributions. Basic and Applied Ecology. 8: 387-397.

Duffy, J. E. 2009. Why biodiversity is important to the functioning of real-world ecosystems. Frontiers in Ecology and the Environment. 7: 437-444.

Dunham, J. B. Personal communication. Research Fisheries Scientist, U.S. Geological Survey, Corvallis, Oregon.

Dunham, J. B.; Chandler, G.; Rieman, B. E.; Martin, D. 2005. Measuring stream temperature with digital dataloggers: a user's guide. Gen. Tech. Rep. RMRS-GTR-150WWW. Fort Collins, CO: U.S. Department of Agriculture, Forest Service, Rocky Mountain Research Station. 15 p.

Dunham, J. B.; Rieman, B. E. 1999. Metapopulation structure of bull trout: influences of physical, biotic, and geometrical landscape characteristics. Ecological Applications. 9: 642-655.

Dunham, J. B.; Rieman, B. E.; Peterson, J. T. 2002. Patch-based models of species occurrence: lessons from salmonid fishes in streams. In: Scott, J. M.; Heglund, P. J.; Morrison, M.; Raphael, M.; Haufler, J.; Wall, B., eds. Predicting species occurrences: issues of scale and accuracy. Covelo, CA: Island Press: 327-334.

Dunham, J. B.; Young, M. K.; Gresswell, R. E.; Rieman, B. E. 2003. Effects of fires on fish populations: landscape perspectives on persistence of native fishes and nonnative fish invasions. Forest Ecology and Management. 178: 183-196

Dunham, Jason B.; Rosenberger, Amanda E.; Luce, Charlie H.; Rieman, Bruce E. 2007. Influences of wildfire and channel reorganization on spatial and temporal variation in stream temperature and the distribution of fish and amphibians. Ecosystems. 10: 335-346.

Eaton, J. G.; McCormick, J. H.; Stefan, H. G.; Hondzo, M. 1995. Extreme value analysis of a fish/temperature field database. Ecological Engineering. 4: 289-305.

Eby, L. A.; Fagan, W. F.; Minckley, W. L. 2003. Variability and dynamics of a desert stream community. Ecological Applications. 13(6): 1566-1579.

Eby, L. A.; Roach, J.; Crowder, L. B.; Stanford, J. A. 2006. Effects of stocking-up freshwater food webs. Trends in Ecology and Evolution. 21: 576-584.

Elliott, J. M.; Hurley, M. A.; Maberly, S. C. 2000. The emergence period of sea trout fry in a Lake District stream correlates with the North Atlantic Oscillation. Journal of Fish Biology. 56: 208-210.

Fagan, W. F. 2002. Connectivity, fragmentation, and extinction risk in dendritic metapopulations. Ecology. 83: 3243-3249.

Falcone, J. A.; Carlisle, D. M.; Wolock, D. M.; Meador, M. R. 2010. GAGES: A stream gage database for evaluating natural and altered flow conditions in the conterminous United States. Ecology. 91: 621.

Fausch, K.; Rieman, B.; Young, M.; Dunham, J. 2006. Strategies for conserving native salmonid populations at risk from nonnative invasions: tradeoffs in using barriers to upstream movement. Gen. Tech. Rep. RMRS-GTR-174. Fort Collins, CO: U.S. Department of Agriculture, Forest Service, Rocky Mountain Research Station. 44 p.

Fausch, K. D. 2008. A paradox of trout invasions in North America. Biological Invasions. 10: 685-701.

Fausch, K. D.; Rieman, B. E.; Dunham, J. B.; Young, M. K.; Peterson, D. P. 2009. Invasion versus isolation: tradeoffs in managing native salmonids with barriers to upstream movement. Conservation Biology. 23: 859-870.

Fausch, K. D.; Taniguchi, Y.; Nakano, S.; Grossman, G. D.; Townsend, C. R. 2001. Flood disturbance regimes influence rainbow trout invasion success. Ecological Applications. 11: 1438-1455.

Ficke, A. D.; Myrick, C. A.; Hansen, L. J. 2007. Potential impacts of global climate change on freshwater fisheries. Reviews in Fish Biology and Fisheries. 17: 581-613.

Finney, B. P.; Gregory-Eaves, I.; Douglas, M. S. V.; Smol, J. P. 2002. Fisheries productivity in the northeastern Pacific Ocean over the past 2,200 years. Nature. 416: 729-733.

Firth, P. L. 1998. Fresh water: perspectives on the integration of research, education, and decision making. Ecological Applications. 8: 601-609.

Flebbe, P. A.; Roghair, L. D.; Bruggink, J. L. 2006. Spatial modeling to project southern Appalachian trout distribution in a warmer climate. Transactions of the American Fisheries Society. 135: 1371-1382.

Francis, R. C.; Sibley, T. H. 1991. Climate change and fisheries: what are the real issues. The Northwest Environmental Journal. 7: 295-307.

Francis, R. I. C. C.; Shotton, R. 1997. Risk in fisheries management: a review. Canadian Journal of Fisheries and Aquatic Sciences. 54: 1699-1715.

Franco, E.; Budy, P. 2004. Linking environmental heterogeneity to the distribution and prevalence of *Myxobolus cerebalis*: a comparison across sites in a northern Utah watershed. Transactions of the American Fisheries Society. 133: 1176-1189.

Frissell, C. A.; Liss, W. J.; Gresswell, R. E.; Nawa, R. K.; Ebersole, J. L. 1997. A resource in crisis: changing the measure of salmon management. In: Stouder, D. J.; Bisson, P. A.; Naiman, N. J., eds. Pacific salmon and their ecosystems: status and future options. NY: Chapman and Hall: 411-446.

Fullerton, A. H.; Steel, E. A.; Caras, Y.; Sheer, M.; Olson, P.; Kaje, J. 2009. Putting watershed restoration in context: alternative future scenarios influence management outcomes. Ecological Applications. 19: 218-235.

Furniss, M. J.; Staab, B. P.; Hazelhurst, S.; Clifton, C. F.; Roby, K. B.; Ilhadrt, B. L.; Larry, E. B.; Todd, A. H.; Reid, L. M.; Hines, S. J.; Bennett, K. A.; Luce, C. H.; Edwards, P. J. 2010. Water, climate change, and forests: watershed stewardship for a changing climate. Gen. Tech. Rep. PNW-GTR-812. Portland, OR: U.S. Department of Agriculture, Forest Service, Pacific Northwest Research Station. 75 p.

Gardner, K. K.; McGlynn, B. L. 2009. Seasonality in spatial variability and influence of land use/land cover and watershed characteristics on stream water nitrate concentrations in a developing watershed in the Rocky Mountain West. Water Resources Research. doi:10.1029/2008WR007029.

Geist, D. R.; Arntzen, E. V.; Murray, C. J.; McGrath, K. E.; Bott, Y. J.; Hanrahan, T. P. 2008. Influence of river level on temperature and hydraulic gradients in chum and fall chinook salmon spawning areas downstream of Bonneville Dam, Columbia River. North American Journal of Fisheries Management. 28: 30-41.

Gibson, S. Y.; Van Der Marel, R. C.; Starzomski, B. M. 2009. Climate change and conservation of leading-edge peripheral populations. Conservation Biology. 23: 1369-1373.

Global Change Research Program [GCRP]. 2009. United States global climate change impacts report. 2009. Karl, T. R.; Melillo, J. M.; Peterson, T. C., eds. Cambridge, UK: Cambridge University Press.

Graumlich, L. J.; Pisaric, M. F. J.; Waggoner, L. A.; Littell, J. S.; King, J. C. 2003. Upper Yellowstone River flow and teleconnections with Pacific basin climate variability during the past three centuries. Climatic Change. 59: 245-262.

Gresswell, R. E. 1999. Fire and aquatic ecosystems in forested biomes of North America. Transactions of the American Fisheries Society. 128(2): 193-221.

Groves, C. R. 2003. Drafting a conservation blueprint: a practitioner's guide to planning for biodiversity. Washington, DC: Island Press.

Guisan, A.; Thuiller, W. 2005. Predicting species distributions: offering more than simple habitat models. Ecology Letters. 8: 993-1009.

Hamlet, A. F.; Lettenmaier, D. P. 1999. Effects of climate change on hydrology and water resources in the Columbia River basin. Journal of the American Water Resources Association. 35: 1597-1623.

Hamlet, A. F.; Lettenmaier, D. P. 2007. Effects of 20th century warming and climate variability on flood risk in the Western US. Water Resources Research. 43: W06427.

Hamlet, A. F.; Mote, P. W.; Clark, M. P.; Lettenmaier, D. P. 2005. Effects of temperature and precipitation variability on snowpack trends in the western United States. Journal of Climate. 18: 4545-4561

Hamlet, A. F.; Mote, P. W.; Clark, M. P.; Lettenmaier, D. P. 2007. Twentieth-century trends in runoff, evapotranspiration, and soil moisture in the western United States. Journal of Climate. 20: 1468-1486.

Hari, R. E.; Livingstone, D. M.; Siber, R.; Burkhardt-Holm, P.; Guttinger, H. 2006. Consequences of climatic change for water temperature and brown trout populations in alpine rivers and streams. Global Change Biology. 12: 10-26.

Harig, A. L.; Fausch, K. D. 2002. Minimum habitat requirements for establishing translocated cutthroat trout populations. Ecological Applications. 12: 535-551.

Harrison, G. W. 1979. Stability under environmental stress: resistance, resilience, persistence, and variability. American Naturalist. 5: 659-669.

Harvey, B. C.; Nakamoto, R. J.; White, J. L. 2006. Reduced streamflow lowers dry-season growth of rainbow trout in a small stream. Transactions of the American Fisheries Society. 135: 998-1005.

Hastings, K. 2005. Long-term persistence of isolated fish populations in the Alexander Archipelago. Missoula, MT: University of Montana. Dissertation.

Hauer, F. R.; Baron, J. S.; Campbell, D. H.; Fausch, K. D.; Hostetler, S. W.; Leavesley, G. H.; Leavitt, P. R.; McKnight, D. M.; Stanford, J. A. 2007. Assessment of climate change and freshwater ecosystems of the Rocky Mountains, USA and Canada. Hydrological Processes. 11: 903-924.

Healey, M. 2009. Resilient salmon, resilient fisheries for British Columbia, Canada. [Online]. Ecology and Society 14(1): 2. Available: http://www.ecologyandsociety.org/vol14/iss11/art12/ [June 2009].

Healey, M. C.; Prince, A. 1995. Scales of variation in life history tactics of Pacific salmon and the conservation of phenotype and genotype. In: Nielsen, J. L., ed. Evolution and the aquatic ecosystem: defining unique units in population conservation. Bethesda, MD: American Fisheries Society. Symposium 17: 176-184.

Heck, M. P. 2007. Effects of wildfire on growth and demographics of coastal cutthroat trout in headwater streams. Corvallis, OR: Oregon State University. Thesis.

Heino, J.; Virkkala, R.; Toivonen, H. 2009. Climate change and freshwater biodiversity: detected patterns, future trends and adaptations in northern regions. Biological Reviews. 84: 39-54.

Hendrickson, S.; Walker, K.; Jacobson, S.; Bower, F. 2008. Assessment of aquatic organism passage at road/stream crossings for the Northern Region of the USDA Forest Service. [Online]. U.S. Department of Agriculture, Forest Service, Northern Region. Available: http://www.fs.fed.us/r1/projects/engineering/fish_passage_web.pdf.

Hendry, A. P.; Wenburg, J. K.; Bentzen, P.; Volk, E. C.; Quinn, T. P. 2000. Rapid evolution of reproductive isolation in the Wild: evidence from introduced salmon. Science. 290: 516-518.

Herlihy, A. T.; Larsen, D. P.; Paulsen, S. G.; Urquhart, N. S.; Rosenbaum, B. J. 2000. Designing a spatially balanced, randomized site selection process for regional stream surveys: the EMAP mid-Atlantic pilot study. Environmental Monitoring and Assessment 63:95-113.

Hessburg, P. F.; Kuhlmann, E. E.; Swetnam, T. W. 2005. Examining the recent climate through the lens of ecology: inferences from temporal pattern analysis. Ecological Applications. 15: 440-457.

Hijmans, Robert J.; Cameron, Susan E.; Parra, Juan I.; Jones, Peter G.; Jarvis, A. 2005. Very high resolution interpolated climate surfaces for global land areas. International Journal of Climatology. 25: 1965-1978.

Hilborn, R.; Quinn, T. P.; Schindler, D. E.; Rogers, D. E. 2003. Biocomplexity and fisheries sustainability. Proceedings of the National Academy of Science. 100: 6564-6568.

Hitt, N. P. 2003. Immediate effects of wildfire on stream temperature. Journal of Freshwater Ecology. 18: 171-173.

Hodgson, J. A.; Thomas, C. D.; Wintle, B. A.; Moilanen, A. 2009. Climate change, connectivity and conservation decision making: back to basics. Journal of Applied Ecology. 46: 964-969.

Hoerling, M.; Eischeid, J. 2007. Past peak water in the Southwest. Southwest Hydrology. 6(1): 18-19, 35.

Holling, C. S.; Meffe, G. K. 1996. Command and control and the pathology of natural resource management. Conservation Biology. 10: 328-337.

Holtby, L. B. 1988. Effects of logging on stream temperatures in Carnation Creek, British Columbia, and associated impacts on the Coho salmon (*Oncorhynchus kisutch*). Canadian Journal of Fisheries and Aquatic Sciences. 45: 502-515.

Honea, J. M.; Jorgensen, J. C.; McClure, M.; Cooney, T. D.; Engie, K.; Holzer, D. M.; Hilborn, R. 2009. Evaluating habitat effects on population status: influence of habitat restoration on spring-run Chinook salmon. Freshwater Biology. 54: 1576-1592.

Hurd, B.; Leary, N.; Jones, R.; Smith, J. 1999. Relative regional vulnerability of water resources to climate change. Journal of the American Water Resources Association. 35: 1399-1409.

Independent Science Advisory Board (ISAB). 2007. Climate change impacts on Columbia River basin fish and wildlife. ISAB Climate Change Report ISAB 2007-2. Portland, OR: Northwest Power and Conservation Council. 136 p.

Intergovernmental Panel on Climate Change (IPCC). 2007. Climate change 2007: the physical science basis. [Online]. Available: http://www.ipcc.ch/ [January 21, 2009].

Isaak, D.; Horan, D.; Wollrab, S. 2010. A simple method using underwater epoxy to permanently install temperature sensors in mountain streams. Available: http://www.fs.fed.us/rm/boise/AWAE/projects/stream_temperature.shtml/ [September 1, 2010].

Isaak, D. J.; Luce, C. H.; Rieman, B. E.; Nagel, D. E.; Peterson, E. E.; Horan, D. L.; Parkes, S.; Chandler, G. L. 2010. Effects of climate change and recent wildfires on stream temperature and thermal habitat for two salmonids in a mountain river network. Ecological Applications. 20(5): 1350-1371. doi: 10.1890/09-0822.

Isaak, D. J.; Rieman, B. E.; Horan, D. 2009. A watershed-scale monitoring protocol for bull trout. Gen. Tech. Rep. GTR-RMRS-224. Fort Collins, CO: U.S. Department of Agriculture, Forest Service, Rocky Mountain Research Station. 25 p.

Isaak, D. J.; Thurow, R. F. 2006. Network-scale spatial and temporal variation in Chinook salmon (*Oncorhynchus tshawytscha*) redd distributions: patterns inferred from spatially continuous replicate surveys. Canadian Journal of Fisheries and Aquatic Sciences. 63: 285-296.

Isaak, D. J.; Thurow, R. F.; Rieman, B. E.; Dunham, J. B. 2007. Relative roles of habitat size, connectivity, and quality on Chinook salmon use of spawning patches. Ecological Applications. 17: 352-364.

Jackson, C. R.; Pringle, C. M. 2010. Ecological benefits of reduced hydrologic connectivity in intensively developed landscapes. BioScience. 60: 37-46.

Jager, H. I.; VanWinkle, W.; Holcomb, B. D. 1999. Would hydrologic climate change in Sierra Nevada streams influence trout persistence? Transactions of the American Fisheries Society. 128: 222-240.

Jentsch, A.; Kreyling, J.; Beierkuhnlein, C. 2007. A new generation of climate-change experiments: events, not trends. Frontiers in Ecology and the Environment. 5: 365-374.

Jonsson, B.; Jonsson, N. 2009. A review of the likely effects of climate change on anadromous Atlantic Salmon *Salmo salar* and brown trout *Salmo trutta*, with particular reference to water temperature and flow. Journal of Fish Biology. 75: 2381-2447.

Jorgensen, J. C.; Honea, J. M.; McClure, M.; Cooney, T. D.; Engie, K.; Holzer, D. M. 2009. Linking landscape-level change to habitat quality: an evaluation of restoration actions on the freshwater habitat of spring-run Chinook salmon. Freshwater Biology. 54: 1560-1575

Joyce, L. A.; Blate, G. M.; Littell, J. S.; McNulty, S. G.; Millar, C. I.; Moser, S. C.; Neilson, R. P.; O'Halloran, K.; Peterson, D. L. 2008. National Forests. Chapter 3. In: Preliminary review of adaptation options for climate-sensitive ecosystems and resources. A report by the U.S. Climate Change Science Program and the Subcommittee on Global Change Research [Julius, S. H.; West, J. M., eds.]. Synthesis and Assessment Products 4.4. Washington, DC: U.S. Environmental Protection Agency: 3-1 to 3-127.

Joyce, L. A.; Blate, G. M.; NcNulty, S. G.; Millar, C. I.; Moser, S.; Neilson, R. P.; Peterson, D. L. 2009. Managing for multiple resources under climate change: National Forests. Environmental Management. 44: 1022-1032.

Juanes, F.; Gephard, S.; Beland, K. F. 2004. Long-term changes in migration timing of adult Atlantic salmon (*Salmo salar*) at the southern edge of the species distribution. Canadian Journal of Fisheries and Aquatic Sciences. 61: 2392-2400.

Kaushal, S. S.; Likens, G. E.; Jaworski, N. A.; Pace, M. L.; Sides, A. M.; Seekell, D.; Belt, K. T.; Secor, D. H.; Wingate, R. L. 2010. Rising stream and river temperatures in the US. Frontiers in Ecology and the Environment. doi:10.1890/090037.

Keefer, M. L.; Peery, C. A.; Heinrich, M. J. 2007. Temperature mediated en route migration mortality and travel rates of endangered Snake River sockeye salmon. Ecology of Freshwater Fish. 17: 136-145.

Keefer, M. L.; Peery, C. A.; High, B. 2009. Behavioral thermoregulation and associated mortality trade-offs in migrating adult steelhead (*Oncorhynchus mykiss*): variability among sympatric populations. Canadian Journal of Fisheries and Aquatic Sciences. 66: 1734-1747.

Keleher, C. J.; Rahel, F. J. 1996. Thermal limits to salmonid distributions in the Rocky Mountain region and potential habitat loss due to global warming: a geographic information system (GIS) approach. Transactions of the American Fisheries Society. 125: 1-13.

Kennedy, B. M.; Peterson, D. P.; Fausch, K. D. 2003. Different life histories of brook trout populations invading mid-elevation and high-elevation cutthroat trout streams in Colorado. Western North American Naturalist. 63: 215-223.

Kirchner, J. W.; Finkel, R. C.; Riebe, C. S.; Granger, D. E.; Clayton, J. L.; King, J. G.; Megahan, W. F. 2001. Mountain erosion over 10 yr, 10 ky, and 10 my time scales. Geology. 29: 591-594.

Knowles, N.; Dettinger, M. D.; Cayan, D. R. 2006. Trends in snowfall versus rainfall in the Western United States. Journal of Climate. 19: 4545-4559.

Labbe, T. R.; Fausch, K. D. 2000. Dynamics of intermitent stream habitat regulate persistence of a threatened fish at multiple scales. Ecological Applications. 10(6): 1774-1791.

Lake, P. S.; Bond, N.; Reich, P. 2007. Linking ecological theory with stream restoration. Freshwater Biology. 52: 597-615.

Latimer, A. M.; Wu, S.; Gelfand, A. E.; Silander, J. A. 2006. Building statistical models to analyze species distributions. Ecological Applications. 16: 33-50.

Lawler, J. J.; Tear, T. H.; Pyke, C.; Shaw, M. R.; Gonzalez, P.; Kareiva, P.; Hansen, L.; Hannah, L.; Klausmeyer K.; Aldous, A.; Bienz, C.; Pearsall, S. 2009. Resource management in a changing and uncertain climate. Frontiers in Ecology and the Environment 7, doi:10.1890/070146.

Lee, D.; Sedell, J.; Rieman, B.; Thurow, R.; Williams, J. 1997. Assessment of the condition of aquatic ecosystems in the Interior Columbia River basin. Chapter 4. Eastside Ecosystem Management Project. PNW-GTR-405. Portland, OR: U.S. Department of Agriculture, Forest Service, Pacific Northwest Research Station.

Lesica, P.; Allendorf, F. W. 1995. When are peripheral populations valuable for conservation? Conservation Biology. 9: 753-760.

Letcher, B. H.; Nislow, K.; Coombs, J.; O'Donnell, M.; Dubreuil, T. 2007. Population response to habitat fragmentation in a stream-dwelling brook trout population. PLoS ONE 2(11): e1139. doi:10.1371/journal.pone.0001139.

Levin, S. A.; Lubchenco, J. 2008. Resilience, robustness, and marine ecosystem-based management. BioScience. 58: 27-32.

Liang, X.; Lettenmaier, D. P. 1994. A simple hydrologically based model of land surface water and energy fluxes for general circulation models. Journal of Geophysical Research. 99: 14415-14428.

Lichatowich, J. 1999. Salmon without rivers: a history of the Pacific salmon crisis. Washington, DC: Island Press.

Littell, J. S.; Mckenzie, D.; Peterson, D. L.; Westerling, A. L. 2009. Climate and wildfire area burned in western U.S. ecoprovinces, 1916-2003. Ecological Applications. 19: 1003-1021.

Logan, J. A.; Powell, J. A. 2001. Ghost forests, global warming, and the mountain pine beetle (*Coleoptera: Scolytidae*). American Entomologist. 47: 160-172.

Luce, C. H.; Holden, Z. A. 2009. Declining annual streamflow distributions in the Pacific Northwest United States. Geophysical Research Letters. 36(L16401): doi:10.1029/2009GL039407.

Ludwig, D.; Hilborn, R.; Walters, C. 1993. Uncertainty, resource exploitation, and conservation: Lessons from history. Science. 260(April): 35-36.

Lüthi, D.; Le Floch, M.; Bereiter, B.; Blunier, T.; Barnola, J. M.; Siegenthaler, U.; Raynaud, D.; Jouzel, J.; Fischer, H.; Kawamura, K.; Stocker, T. F. 2008. High-resolution carbon dioxide concentration record 650,000-800,000 years before present. Nature. 453(7193): 379-382.

Mantua, N.; Hare, S.; Zhang, Y.; Wallace, J. M.; Francis, R. 1997. A Pacific interdecadal climate oscillation with impacts on salmon production. Bulletin of the American Meteorologic Society. 78: 1069-1079.

Mantua, N., unpublished data. Monthly values of the Pacific Decadal Oscillation. Available: http://jisao.washington.edu/pdo/ [September 1, 2010].

Mantua, N.; Toliver, I.; Hamlet, A. 2009. Impacts of climate change on key aspects of freshwater salmon habitat in Washington State. [Online]. Washington State Climate Impacts Group. Available: http://cses.washington.edu/cig/res/ia/waccia.shtml.

Marcot, B. G.; Holthausen, R. S.; Raphael, M. G.; Rowland, M. M.; Wisdom, M. J. 2001. Using bayesian belief networks to evaluate fish and wildlife population viability under land managment alternatives from and environmental impact statement. Forest Ecology and Management. 153: 29-42.

Marks, D.; Kimball, J.; Tingey, D.; Link, T. 1998. The sensitivity of snowmelt processes to climate conditions and forest cover during rain-on-snow: a case study of the 1996 Pacific Northwest flood. Hydrological Processes. 12: 1569-1587.

Mawdsley, J. R.; O'Malley, R.; Ojima, D. S. 2009. A review of climate-change adaptation strategies for wildlife management and biodiversity conservation. Conservation Biology. 23: 1080-1089.

McCabe, G. J.; Palecki, M. A.; Betancourt, J. L. 2004. Pacific and Atlantic Ocean influences on multidecadal drought frequency in the United States. Proceedings National Academy of Sciences. 101: 4136-4141.

McCarthy, S. G.; Duda, J. J.; Emlen, J. M.; Hodgson, G. R.; Beauchamp, D. A. 2009. Linking habitat quality with trophic performance of steelhead along forest gradients in the south fork Trinity River Watershed, California. Transactions of the American Fisheries Society. 138: 506-521.

McClure, M. M.; Carlson, S. M.; Beechie, T.; Pess, G. R.; Jorgensen, J. C.; Sogard, S. M.; Sultan, S. E.; Holzer, D. M.; Travis, J.; Sanderson, B. L.; Power, M. E.; Carmichael, R. W. 2008. Evolutionary consequences of habitat loss for Pacific anadromous salmonids. Evolutionary Applications. 2008: 300-318.

McCullough, D. A.; Bartholow, J. M.; Jager, H. I.; Beschta, R. L.; Cheslak, E. F.; Deas, M. L.; Ebersole, J. L.; Foott, J. S.; Johnson, S. L.; Marine, K. R.; Mesa, M. G.; Petersen, J. H.; Souchon, Y.; Tiffan, K. F.; Wurtsbaugh, W. A. 2009. Research in thermal biology: burning questions for coldwater stream fishes. Reviews in Fisheries Science. 17: 90-115.

McGrath, K. E.; Scott, J. M.; Rieman, B. E. 2008. Length variation in age-0 westslope cutthroat trout at multiple spatial scales. North American Journal of Fisheries Management. 28: 1529-1540

McIntire, E. J. B.; Fajardo, A. 2009. Beyond description: the active and effective way to infer processes from spatial patterns. Ecology. 90: 46-56.

McKenzie, D.; Gedalof, Z.; Peterson, D. L.; Mote, P. 2004. Climate change, wildfire, and conservation. Conservation Biology. 18: 890-902.

McPhaden, M. J.; Busalacchi, A. J.; Cheney, R.; Donguy, J. R.; Gage, K. S.; Halpern, D.; Ji, M.; Julian, P.; Meyers, G.; Mitchum, G. T.; Niiler, P. P.; Picaut, J.; Reynolds, R. W.; Smith, N.; Takeuchi, K. 1998. The Tropical Ocean-Global Atmosphere observing system: a decade of progress. Journal of Geophysical Research. 103: 14169-14240.

McPhail, J. D.; Lindsey, C. C. 1986. Zoogeography of the freshwater fishes of Cascadia (the Columbia system and rivers north to the Stikine). In: Hocutt, C. H.; Wiley, E. O., eds. The zoogeography of North American freshwater fishes. New York: John Wiley and Sons: 615-638.

Meehl, G. A.; Tebaldi, C.; Walton, G.; Easterling, D.; McDaniel, L. 2009. Relative increase of record high maximum temperatures compared to record low minimum tempcratures in the U.S. Geophysical Research Letters. 36 (L23701): doi:10.1029/2009GL040736.

Meier, W.; Bonjour, C.; Wüest, A.; Reichert, P. 2003. Modeling the effect of water diversion on the temperature of mountain streams. Journal of Environmental Engineering. 129: 755-764.

Meyer, K. A.; Elle, F. S.; Lamansky, J. A., Jr. 2009. Environmental factors related to the distribution, abundance, and life history characteristics of mountain whitefish in Idaho. North American Journal of Fisheries Management. 29: 753-767.

Meyer, K. A.; Lamansky, J. A.; Schill, D. J. 2006. Evaluation of an unsuccessful brook trout electrofishing removal project in a small Rocky Mountain stream. North American Journal of Fisheries Management. 26: 849-860.

Millar, C. I.; Stephenson, N. L.; Stephens, S. L. 2007. Climate change and forests of the future: managing in the face of uncertainty. Ecological Applications. 17: 2145-2151.

Miller, D.; Luce, C.; Benda, L. 2003. Time, space, and episodicity of physical disturbance in streams. Forest Ecology and Management. 178: 89-104.

Milner, A. M. 1987. Colonization and ecological development of new streams in Glacier Bay National Park, Alaska. Freshwater Biology. 18: 53-70.

Milner, A. M.; Knudsen, E. E.; Soiseth, C.; Robertson, A. L.; Schell, D.; Phillips, I. T.; Magnusson, K. 2000. Colonization and development of stream communities across a 200-year gradient in Glacier Bay National Park, Alaska, USA. Canadian Journal of Fisheries and Aquatic Sciences. 57: 2319-2335.

Milner, A. M.; Robertson, A. L.; Monaghan, K. A.; Veal, A. J.; Flory, E. A. 2008. Colonization and development of an Alaskan stream community over 28 years. Frontiers in Ecology and the Environment. 6: 413-419.

Minckley, W. L.; Deacon, J. E. 1991. Battle against extinction: native fish management in the American West. Tucson, AZ: University of Arizona Press.

Mohseni, O.; Stefan, H. G. 1999. Stream temperature/air temperature relationship: a physical interpretation. Journal of Hydrology. 218: 128-141.

Mohseni, O.; Erickson, T. R.; Stefan, H. G. 1999. Sensitivity of stream temperatures in the United States to air temperatures projected under a global warming scenario. Water Resources Research. 35: 3723-3733.

Mohseni, O.; Stefan, H. G.; Eaton, J. G. 2003. Global warming and potential changes in fish habitat in U.S. streams. Climatic Change. 59: 389-409.

Montgomery, D. R. 2000. Coevolution of the Pacific salmon and Pacific Rim topography. Geology. 28: 1107-1110.

Montgomery, D. R.; Beamer, E. M.; Pess, G. R.; Quinn, T. P. 1999. Channel type and salmonid spawning distribution and abundance. Canadian Journal of Fisheries and Aquatic Sciences. 56: 377-387.

Moore, M. V.; Hampton, S. E.; Izmest'eva, L. R.; Silow, E. A.; Peshkova, E. V.; Pavlov, B. K. 2009. Climate change and the world's sacred sea-Lake Baikal, Siberia. BioScience. 59: 405-417.

Morgan, P.; Heyerdahl, E. K.; Gibson, C. E. 2008. Multiseason climate synchronized widespread forest fires throughout the 20th-century, Northern Rockies, USA. Ecology. 89: 717-728.

Morita, K.; Yamamoto, S. 2002. Effect of habitat fragmentation by damming on the persistence of stream-dwelling charr populations. Conservation Biology. 16: 1318-1323.

Morrill, J. C.; Bales, R. C.; Asce, M.; Conklin, M. H. 2005. Estimating stream temperature from air temperature: implications for future water quality. Journal of Environmental Engineering. 131: 139-146.

Morrison, J.; Quick, M. C.; Foreman, M. G. G. 2002. Climate change in the Fraser River watershed: flow and temperature projections. Journal of Hydrology. 263: 230-244.

Mote, P. W.; Hamlet, A. F.; Clark, M. P.; Lettenmaier, D. P. 2005. Declining mountain snowpack in western North America. Bulletin of the American Meteorological Society. 86: 39-49.

Mote, P. W.; Parson, E. A.; Hamlet, A. F.; Keeton, W. S.; Lettenmaier, D.; Mantua, N.; Miles, E. L.; Peterson, D. W.; Peterson, D. L.; Slaughter, R.; Snover, A. K. 2003. Preparing for climatic change: the water, salmon, and forests of the Pacific Northwest. Climatic Change. 61: 45-88.

Mote, P. W.; Salathé, E.; Dulière, V.; Jump, E. 2008. Scenarios of future climate for the Pacific Northwest. Report prepared by the Climate Impacts Group, Center for Science in the Earth System, Joint Institute for the Study of the Atmosphere and Oceans. Seattle, WA: University of Washington.

Muhlfeld, C. C.; McMahon, T. E.; Boyer, M. C.; Gresswell, R. E. 2009. Local habitat, watershed, and biotic factors influencing the spread of hybridization between native westslope cutthroat trout and introduced rainbow trout. Transactions of the American Fisheries Society. 138: 1036-1051.

Naiman, R. J.; Magnuson, J. J.; Firth, P. L. 1998. Integrating cultural, economic, and environmental requirements for fresh water. Ecological Applications. 8(3): 569-570.

Nakano, S.; Kitano, F.; Maekawa, K. 1996. Potential fragmentation and loss of thermal habitats for charrs in the Japanese archipelago due to climatic warming. Freshwater Biology. 36: 711-722.

NASA Goddard Institute for Space Studies. January to December global mean temperature over land and ocean. Available: http://data.giss.nasa.gov/gistemp/ [September 1, 2010].

Nelitz, M.; Porter, M.; Bennett, K.; Werner, A.; Bryan, K.; Poulsen, F.; Carr, D. 2009. Evaluating the vulnerability of freshwater fish habitats to the effects of climate change in the Carboo-Chilcotin. [Online]. Vancouver, BC: ESSA Technologies Ltd. Available: http://www.thinksalmon.com/fswp_project/item/evaluating_the_vulnerability_of_pacific_salmon_to_effects_of_climate_change/ [March 14, 2009].

Neraas, L. P.; Spruell, P. 2001. Fragmentation of riverine systems: the origins of bull trout (Salvelinus confuentus) collected at the base of Cabinet Gorge Dam, Montana. Molecular Ecology. 10: 1153-1164

Neuheimer, A. B.; Taggart, C. T. 2007. The growing degree-day and fish size-at-age: the overlooked metric. Canadian Journal of Fisheries and Aquatic Sciences. 64: 375-385.

Neville, H.; Dunham, J.; Peacock, M. 2006. Assessing connectivity in salmonid fishes with DNA microsatellite markers. In: Crooks, K. R.; Sankayan, M., eds. Connectivity conservation. Cambridge, UK: Cambridge University Press: 318-342.

Neville, H.; Dunham, J. B.; Rosenberger, A.; Umek, J.; Nelson, B. 2009. Influences of wildfire, habitat size, and connectivity on trout in headwater streams revealed by patterns of genetic diversity. Transactions of the American Fisheries Society. 138: 1314-1327

Northcote, T. G. 1997. Potamodromy in salmonidae—living and moving in the fast lane. North American Journal of Fisheries Management. 17: 1029-1045.

Noss, R. F. 2001. Beyond Kyoto: forest management in a time of rapid climate change. Conservation Biology. 15: 578-590.

Noss, R. F.; Beier, P.; Covington, W. W.; Grumbine, R. E.; Lindenmayer, D. B.; Prather, J. W.; Schmiegelow, F.; Sisk, T. D.; Vosick, D. J. 2006. Recommendations for integrating restoration ecology and conservation biology in ponderosa pine forests of the southwestern United States. Restoration Ecology. 14: 4-10.

Olden, J. D.; Naiman, R. J. 2009. Incorporating thermal regimes into environmental assessments: modifying dam operations to restore freshwater ecosystem integrity. Freshwater Biology. doi:10.1111/j.1365-2427.2009.02179.x.

Pagano, T.; Garen, D. 2005. A recent increase in western U.S. streamflow variability and persistence. Journal of Hydrometeorology. 6: 173-179.

Parmesan, C.; Yohe, G. 2003. A globally coherent fingerprint of climate change impacts across natural systems. Nature. 421: 37-42.

Paul, A. J.; Post, J. R. 2001. Spatial distribution of native and non-native salmonids in streams of the eastern slopes of the Canadian Rocky Mountains. Transactions of the American Fisheries Society. 130: 417-430.

Pederson, G. T.; Graumlich, L. J.; Fagre, D. B.; Kipfer, T.; Muhlfeld, C. C. 2009. A century of climate and ecosystem change in Western Montana: what do temperature trends portend? Climatic Change. 96: doi 10.1007/s10584-009-9642-y.

Petersen, J. H.; Kitchell, J. F. 2001. Climate regimes and water temperature changes in the Columbia River: bioenergetic implications for predators of juvenile salmon. Canadian Journal of Fisheries and Aquatic Sciences. 58: 1831-1841.

Peterson, D. P.; Fausch, K. D.; Watmough, J.; Cunjak, R. A. 2008. When eradication is not an option: modeling strategies for electrofishing suppression of nonnative brook trout to foster persistence of sympatric native cutthroat trout in small streams. North American Journal of Fisheries Management. 28: 1847-1867.

Peterson, D. P.; Rieman, B. E.; Young, M. K.; Brammer, J. A. 2010. Modeling predicts that redd trampling by cattle may contribute to population declines of native trout. Ecological Applications. 20(4): 954-966.

Peterson, E. E.; Theobald, D. M.; Ver Hoef, J. M. 2007. Geostatistical modeling on stream networks: developing valid covariance matrices based on hydrologic distance and stream flow. Freshwater Biology. 52: 267-279.

Peterson, E. E.; Urquhart, N. S. 2006. Predicting water quality impaired stream segments using landscape-scale data and a regional geostatistical model: a case study in Maryland. Environmental Monitoring and Assessment. 121: 615-638.

Peterson, E. E.; Ver Hoef, J. M. 2010. A mixed-model moving-average approach to geostatistical modeling in stream networks. Ecology. 91: 644-651.

Pettit, N. E.; Naiman, R. J. 2007. Fire in the riparian zone: characteristics and ecological consequences. Ecosystems. 10: 673-687.

Pielke, R. 2009. Collateral damage from the death of stationarity. GWEX News (May): 5-7.

Pittock, A. B. 2006. Are scientists underestimating climate change? EOS. 87: 340-341.

Poff, N. L.; Angermeier, P. L.; Cooper, S. D.; Lake, P. S.; Fausch, K. D.; Winemiller, K. O.; Mertes, L. A. K.; Oswood, M. W.; Reynolds, J.; Rahel, F. J. 2001. Fish diversity in streams and rivers. In: Chapin, F.S. III; Sala, O.E.; Huber-Saanwald, E., eds. Global biodiversity in a changing environment: scenarios for the 21st century. New York: Springer-Verlag: 315-349.

Poff, N. L.; Brinson, M. M.; Day, J. W. J. 2002. Aquatic ecosystems and global climate change: potential impacts on inland freshwater and coastal wetland ecosystems in the United States. [Online]. Report from the Pew Center on Global Climate Change. Available: http:www.pewclimate.org/projects/aquatic.

Poff, N. L.; Richter, B. D.; Arthington, A. H.; Bunn, S. E.; Naiman, R. J.; Kendy, E.; Acreman, M.; Apse, C.; Bledsoe, B.; Freeman, M. C.; Henriksen, J.; Jacobson, R. B.; Kennen, J. G.; Merritt, D. M.; O'Keeffe, J. H.; Olden, J.; Rogers, K.; Tharme, R. E.; Warner, A. 2010. The ecological limits of hydrologic alteration (ELOHA): a new framework for developing regional environmental flow standards. Freshwater Biology. 55: 147-170.

Pont, D.; Hughes, R. M.; Whittier, T. R.; Schmutz, S. 2009. A predictive index of biotic integrity model for aquatic-vertebrate assemblages of western U.S. streams. Transactions of the American Fisheries Society. 138: 292-305.

Poole, G. C.; Dunham, J. B.; Keenan, D. M.; Sauter, S. T.; McCullough, D. A.; Mebane, C.; Lockwood, J. C.; Essig, D. A.; Hicks, M. P.; Sturdevant, D. J.; Materna, E. J.; Spalding, S. A.; Risley, J.; Deppman, M. 2004. The case for regime-based water quality standards. BioScience. 54: 155-161

Pörtner, H. O.; Farrell, A. P. 2008. Physiology and climate change. Science. 322: 690-692.

Propst, D. L.; Gido, K. B.; Stefferud, J. A. 2008. Natural flow regimes, nonnative fishes, and native fish persistence in arid-land river systems. Ecological Applications. 18: 1236-1252.

Quinn, T. P. 2005. The behavior and ecology of Pacific salmon and trout. Seattle, WA: University of Washington Press.

Quinn, T. P.; Adams, D. J. 1996. Environmental changes affecting the migratory timing of American Shad and Sockeye salmon. Ecology. 77: 1151-1162.

Quinn, T. P.; Hodgson, S.; Peven, C. 1997. Temperature, flow, and the migration of adult sockeye salmon (*Oncorhynchus nerka*) in the Columbia River. Canadian Journal of Fisheries and Aquatic Sciences. 54: 1349-1360.

Rahel, F. J.; Bierwagen, B.; Taniguchi, Y. 2008. Managing aquatic species of conservation concern in the face of climate change and invasive species. Conservation Biology. 22: 551-561.

Rahel, F. J.; Keleher, C. J.; Anderson, J. L. 1996. Potential habitat loss and population fragmentation for cold water fish in the North Platte River drainage of the Rocky Mountains: response to climate warming. Limnology and Oceanography. 41: 116-1123.

Raupach, M. R.; Marland, G.; Ciais, P.; Le Que′re′, C.; Canadell, J. G.; Klepper, G.; Field, C. B. 2007. Global and regional drivers of accelerating CO_2 emissions. Proceedings of the National Academy of Sciences. 104: 10288-10293.

Reckhow, K. H. 1999. Water quality prediction and probability network models. Canadian Journal of Fisheries and Aquatic Sciences. 56(7): 1150-1158.

Reese, C. D.; Harvey, B. C. 2002. Temperature-dependent interactions between juvenile steelhead and Sacramento pikeminnow in laboratory streams. Transactions of the American Fisheries Society. 131: 599-606.

Reeves, G. H.; Benda, L. E.; Burnett, K. M.; Bisson, P. A.; Sedell, J. R. 1995. A disturbance-based ecosystem approach to maintaining and restoring freshwater habitats of evolutionarily significant units of anadromous salmonids in the Pacific Northwest. In: Nielsen, J. Evolution and the aquatic ecosystem. Bethesda, MD: Proceedings of the 17th Symposium of the American Fisheries Society: 334-349.

Reeves, G. H.; Bisson, P. A.; Dambacher, J. M. 1998. Fish communities. In: Naiman, R.J.; Bilby, R.E., eds. River ecology and management: lessons from the Pacific coastal ecoregion. Springer-Verlag: 200-234.

Regonda, S. K.; Rajagopalan, B.; Clark, M.; Pitlick, J. 2005. Seasonal cycle shifts in hydroclimatology over the Western United States. Journal of Climate. 18: 372-384.

Rhodes, J. J.; Baker, W. L. 2008. Fire probability, fuel treatment effectiveness and ecological tradeoffs in western U.S. public forests. The Open Forest Science Journal. 1: 1-7.

Ricciardi, A.; Simberloff, D. 2009. Assisted colonization is not a viable conservation strategy. Trends in Ecology and Evolution. 24: 248-253.

Richter, A.; Kolmes, S. A. 2005. Maximum temperature limits for Chinook, Coho, and chum salmon, and steelhead trout in the Pacific Northwest. Reviews in Fisheries Science. 13: 23-49.

Rieman, B. E., Allendorf, F. W. 2001. Effective population size and genetic conservation criteria for bull trout. North American Journal of Fisheries Management. 21: 756-764.

Rieman, B. E.; Clayton, J. 1997. Fire and fish: issues of forest health and conservation of native fishes. Fisheries. 22: 6-15.

Rieman, B. E.; Dunham, J. B. 2000. Metapopulation and salmonids: a synthesis of life history patterns and empirical observations. Ecology of Freshwater Fish. 9: 51-64.

Rieman, B. E.; Isaak, D. J. Unpublished data. Summer stream temperature data from 1993 to 2006 for five headwater streams in central Idaho.

Rieman, B. E.; Hessburg, P. F.; Luce, C.; Dare, M. R. 2010. Wildfire and management of forests and native fishes: conflict or opportunity for convergent solutions. BioScience. 60: 460-468.

Rieman, B. E.; Isaak, D.; Adams, S.; Horan, D.; Nagel, D.; Luce, C. 2007. Anticipated climate warming effects on bull trout habitats and populations across the Interior Columbia River basin. Transactions of the American Fisheries Society. 136: 1552-1565

Rieman, B. E.; Lee, D.; Burns, D.; Gresswell, R.; Young, M.; Stowell, R.; Howell, P. 2003. Status of native fishes in the Western United States and issues for fire and fuels management. Forest Ecology and Management. 178(1-2): 19-212.

Rieman, B. E.; Lee, D. C.; Thurow, R. F.; Hessburg P. F.; Sedell, J. R. 2000. Toward an integrated classification of ecosystems: defining opportunities for managing fish and forest health. Environmental Management. 25(4): 425-444.

Rieman, B. E., McIntyre, J. D. 1995. Occurrence of bull trout in naturally fragmented habitat patches of varied size. Transactions of the American Fisheries Society. 124: 285-296.

Robards, M. D.; Quinn, T. P. 2002. The migratory timing of adult summer-run steelhead in the Columbia River over six decades of environmental change. Transactions of the American Fisheries Society. 131: 523-536.

Roni, P.; Beechie, T. J.; Bilby, R. E.; Leonetti, F. E.; Pollock, M. M.; Pess, G. R. 2002. A review of stream restoration techniques and a hierarchical strategy for prioritizing restoration in Pacific Northwest Watersheds. North American Journal of Fisheries Management. 22: 1-20.

Rood, S. B.; Pan, J.; Gill, K. M.; Franks, C. G.; Samuelson, G. M.; Shepherd, A. 2008. Declining summer flows of Rocky Mountain rivers: changing seasonal hydrology and probably impacts on floodplain forests. Journal of Hydrology. 349: 397-410.

Root, T. L.; Price, J. T.; Hall, K. R.; Schneider, S. H.; Rosenweig, C.; Pounds, J. A. 2003. Fingerprints of global warming on wild animals and plants. Nature. 421: 57-60.

Roper, B. B.; Buffington, J. M.; Bennett, S.; Lanigan, S. H.; Archer, E.; Downie, S. T.; Faustini, J.; Hillman, T. W.; Hubler, S.; Jones, K.; Jordan, C.; Kaufmann, P. R.; Merritt, G.; Moyer, C.; Pleus, A. 2010. A comparison of the performance and compatibility of protocols used by seven monitoring groups to measure stream habitat in the Pacific Northwest. North American Journal of Fisheries Management. 30: 565-587.

Santhi, C.; Allen, P. M.; Muttiah, R. S.; Arnold, J. G.; Tuppad, P. 2008. Regional estimation of base flow for the conterminous United States by hydrologic landscape regions. Journal of Hydrology. 351: 139-153.

Saunders, S.; Montgomery, C.; Easley, T.; Spencer, T. 2008. Hotter and drier: the West's changed climate. [Online]. The Rocky Mountain Climate Organization. Available: http://www.rockymountainclimate.org/ [March, 2008].

Sax, D. F.; Smith, K. F.; Thompson, A. R. 2009. Managed relocation: a nuanced evaluation is needed. Trends in Ecology and Evolution. 24: 472-473.

Schindler, D. E.; Augerot, X.; Fleishman, E.; Mantua, N. J.; Riddell, B.; Ruckelshaus, M.; Seeb, J.; Webster, M. 2008. Climate change, ecosystem impacts, and management for Pacific Salmon. Fisheries. 33: 502-506.

Schneider, P.; Hook, S. J.; Radocinski, R. G.; Corlett, G. K.; Hulley, G. C.; Schladow, S. G.; Steissberg, T. E. 2009. Satellite observations indicate rapid warming trend for lakes in California and Nevada. Geophysical Research Letters. 36(L22402).

Schumm, S. A.; Lichty, R. W. 1965. Time, space, and causality in geomorphology. American Journal of Science. 263: 110-119.

Scott, J. M.; Csuti, B. 1997. Noah worked two jobs. Conservation Biology. 11: 1255-1257.

Selong, J. H.; McMahon, T. E.; Zale, A. V.; Barrows, F. T. 2001. Effect of temperature on growth and survival of bull trout, with application of an improved method for determining thermal tolerance in fishes. Transactions of the American Fisheries Society. 130: 1026-1037.

Shepard, B. B. 2002. A native westslope cutthroat trout population responds positively after brook trout removal and habitat restoration. Intermountain Journal of Sciences. 8: 193-214.

Shuter, B. J.; Post, J. R. 1990. Climate, population variability, and the zoogeography of temperate fishes. Transactions of the American Fisheries Society. 119: 314-336.

Solomon, S.; Plattnerb, G.; Knuttic, R.; Friedlingsteind, P. 2009. Irreversible climate change due to carbon dioxide emissions. Proceedings of the National Academy of Sciences. 106: 1704-1709.

Stewart, I. T.; Cayan, D. R.; Dettinger, M. D. 2005. Changes toward earlier streamflow timing across western North America. Journal of Climate. 18: 1136-1155.

Stockwell, C. A.; Hendry, A. P.; Kinnison, M. T. 2003. Contemporary evolution meets conservation biology. Trends in Ecology and Evolution. 18: 94-101.

Tague, C.; Grant, G.; Farrell, M.; Choate, J.; Jefferson, A. 2008. Deep groundwater mediates streamflow response to climate warming in the Oregon Cascades. Climatic Change. 86: 189-210.

Thurow, R. F. Personal communication. Research Fisheries Scientist, U.S. Forest Service, Rocky Mountain Research Station, Boise, Idaho.

Tiffan, K. F.; Kock, T. J.; Connor, W. P.; Steinhorst, R. K.; Rondorf, D. W. 2009. Behavioral thermal regulation by subyearling fall (autumn) Chinook salmon *Onchorhynchus tshawytscha* in a reservoir. Journal of Fish Biology. 74: 1562-1579.

USDA Forest Service. 2008. Forest Service strategic framework for responding to climate change. [Online]. U.S. Department of Agriculture, Forest Service. Available: http://www.fs.fed.us/climatechange/documents/strategic-framework-climate-change-1-0.pdf [October 8, 2008].

Van Kirk, R. W.; Benjamin, L. 2001. Status and conservation of salmonids in relation to hydrologic integrity in the Greater Yellowstone Ecosystem. Western North American Naturalist. 61: 359-374.

van Mantgem, P. J.; Stephenson, N. L. 2007. Apparent climatically induced increase of tree mortality rates in a temperate forest. Ecology Letters. 10: 909-916.

van Mantgem, P. J.; Stephenson, N. L.; Byrne, J. C.; Daniels, L. D.; Franklin, J. F.; Fule, P. Z.; Harmon, M. E.; Larson, A. J.; Smith, J. M.; Taylor, A. H.; Veblen, T. T. 2009. Widespread increase of tree mortality rates in the Western United States. Science. 323: 521-524.

Ver Hoef, J. M.; Peterson, E. E. 2010. A moving average approach for spatial statistical models of stream networks. Journal of the American Statistical Association. 105: 6-18.

Vitt, P.; Havens, K.; Hoegh-Guldberg, O. 2009. Assisted migration: part of an integrated conservation strategy. Trends in Ecology and Evolution. 24: 473-474.

Volk, E. C.; Bottom, D. L.; Jones, K. K.; Simenstad, C. A. 2010. Reconstructing juvenile chinook salmon life history in the Salmon River estuary, Oregon, using otolith microchemistry and microstructure. Transactions of the American Fisheries Society. 139: 535-549.

Waples, R.; Beechie, T.; Pess, G. R. 2009. Evolutionary history, habitat disturbance regimes, and anthropogenic changes: what do these mean for resilience of Pacific Salmon populations? [Online]. Ecology and Society. 14: 3. Available: http://www.ecologyandsociety.org/vol14/iss11/art13/.

Waples, R. S.; Pess, G. R.; Beechie, T. 2008. Evolutionary history of Pacific salmon in dynamic environments. Evolutionary Applications. 1(2): 189-206.

Waples, R. S.; Zabel, R. W.; Scheuerell, M. D.; Sanderson, B. L. 2007. Evolutionary response by native species to major anthropogenic changes to their ecosystems: Pacific salmon in the Columbia River hydropower systems. Molecular Ecology. 17: 84-96.

Wara, M. W.; Ravelo, A. C.; Delaney, M. L. 2005. Permanent El Niño-like conditions during the Pliocene warm period. Science. 309: 758-761.

Wehner, M. 2005. Changes in daily precipitation and surface air temperature extremes in the IPCC AR4 models. US CLIVAR Variations. 3(3): 5-9.

Welch, D. 2005. What should protected areas managers do in the face of climate change? The George Wright Forum. 22: 75-93.

Wenger, S. J.; Luce, C. H.; Hamlet, A. F.; Isaak, D. J.; Neville, H. M. 2010. Macroscale hydrologic modeling of ecologically relevant flow metrics. Water Resources Research. doi:10.1029/2009WR008839.

West, J. M.; Salm, R. V. 2003. Resistance and resilience to coral bleaching: implications for coral reef conservation and management. Conservation Biology. 17: 956-967.

Westerling, A. L.; Hidalgo, H. G.; Cayan, D. R.; Swetnam, T. W. 2006. Warming and earlier spring increases western U.S. forest wildfire activity. Science. 313: 940-943.

Wiens, J. A.; Bachelet, D. 2009. Matching the multiple scales of conservation with the multiple scales of climate change. Conservation Biology. 24: 51-62.

Williams, J. G.; Zabel, R. W.; Waples, R.; Hutchings, J. A.; Connor, W. P. 2008. Potential for anthropogenic disturbances to influence evolutionary change in the life history of a threatened salmonid. Evolutionary Applications. 2008: 271-285.

Williams, J. W.; Jackson, S. T. 2008. Novel climates, no-analog communities, and ecological surprises. Frontiers in Ecology and the Environment. 5: 475-482.

Williams, J. W.; Jackson, S. T.; Kutzbach, J. E. 2007. Projected distributions of novel and disappearing climates by 2100 AD. Proceedings of the National Academy of Sciences. 104: 5738-5742.

Wolock, D. M.; Winter, T. C.; McMahon, G. 2004. Delineation and evaluation of hydrologic-landscape regions in the United States using geographic information system tools and multivariate statistical analyses. Environmental Management. 34: S71-S88.

Wood, C. C.; Bickham, J. W.; Nelson, R. J.; Foote, C. J.; Patton, J. C. 2008. Recurrent evolution of life history ecotypes in sockeye salmon: implications for conservation and future evolution. Evolutionary Applications. 2: 207-221.

Wooldridge, S. A.; Done, T. J. 2009. Improved water quality can ameliorate effects of climate change on corals. Ecological Applications. 19: 1492-1499.

Wrona, F. J.; Prowse, T. D.; Reist, J. D.; Hobbie, J. E.; Levesque, L. M. J.; Vincent, W. F. 2006. Climate change effects on aquatic biota, ecosystem structure and function. Ambio. 35: 359-369.

Yarnell, S. M.; Viers, J. H.; Mount, J. F. 2010. Ecology and management of the spring snowmelt recession. BioScience. 60: 114-127.

Yeh, Sang-Wook; Kug, Jong-Seong; Dewitte, Boris; Kwon, Min-Ho; Kirtman, Ben P.; Jin, Fei-Fei. 2009. El Niño in a changing climate. Nature. 461: 511-514.

Young, M. K. 2008. Colorado River cutthroat trout: a technical conservation assessment. Gen. Tech. Rep. RMRS-GTR-207WWW. Fort Collins, CO: U.S. Department of Agriculture, Forest Service, Rocky Mountain Station. 123 p.

Zoellick, B. W. 1999. Stream temperatures and the elevational distribution of redband trout in southwestern Idaho. Great Basin Naturalist. 59: 136-143.

The Rocky Mountain Research Station develops scientific information and technology to improve management, protection, and use of the forests and rangelands. Research is designed to meet the needs of the National Forest managers, Federal and State agencies, public and private organizations, academic institutions, industry, and individuals. Studies accelerate solutions to problems involving ecosystems, range, forests, water, recreation, fire, resource inventory, land reclamation, community sustainability, forest engineering technology, multiple use economics, wildlife and fish habitat, and forest insects and diseases. Studies are conducted cooperatively, and applications may be found worldwide.

Station Headquarters
Rocky Mountain Research Station
240 W Prospect Road
Fort Collins, CO 80526
(970) 498-1100

Research Locations

Flagstaff, Arizona	Reno, Nevada
Fort Collins, Colorado	Albuquerque, New Mexico
Boise, Idaho	Rapid City, South Dakota
Moscow, Idaho	Logan, Utah
Bozeman, Montana	Ogden, Utah
Missoula, Montana	Provo, Utah

www.ingramcontent.com/pod-product-compliance
Lightning Source LLC
Chambersburg PA
CBHW080649180526
45168CB00008B/3356